Agricultural Process Engineering

Dr. R.N. Reddy is a product of College of Agricultural Engineering, Baptala. He is engaged in teaching the UG Courses in the field of agricultural process engineering. He worked with various institutes and has participated in many national and international conferences. His 23 articles are published in many national and international reputed journals. As prolific writer he authored 2 more books on Agricultural Engineering.

Agricultural Process Engineering

Editor

R.N. Reddy

2025

Gene-Tech Books

ISBN 81-89729-97-7
ISBN 978-81-89729-97-4

Published by : **GENE-TECH BOOKS**
4762-63/23, Ansari Road, Darya Ganj,
NEW DELHI - 110 002
Phone: 011-4300 3222
e-mail: genetechbooks@yahoo.co.in

PRINTED IN INDIA

Preface

As the world population grows, the productivity of the world's cropland must increase rapidly maintain the minimal supplies currently available in the world. Keeping such a necessity as primary in its agenda, agricultural process engineering involves the application of engineering ad scientific principles to agricultural production, farm management and management of natural resources. As agriculture becomes increasingly mechanised, it becomes pertinent to consider not just the proper application of the best engineering technology to agriculture, but also its relevance, and agricultural process engineering provides a step in the right direction.

The current text aims to delineate for the readers about what entails agricultural process engineering. Discussing its principles, concepts and practices, the book seeks to acquaint the reader with the primary concerns of this field which includes such practices as designing agricultural machinery, power system design, environmental science, food and bioprocessing technology soil and water conservation, etc. The aim of the text is to not just list and discuss these endeavours, but to infact make assessment of the opportunities, challenges and prospects offered by it. In addition, it vies coverage to current trends and developments in the field, lending its contents a more relevant outlook.

R.N. Reddy

Contents

	Preface	*v*
1.	Introduction to Agro-Processing	1
2.	Seed Processing Techniques	31
3.	Processing of Horticultural Crops	67
4.	Fruit Specific Processing Technologies	77
5.	Vegetable Specific Processing Technologies	113
6.	Coffee Processing	153
7.	Freezing of Fruits and Vegetables	171
8.	Dairy Process Engineering	205
9.	Engineering Properties of Foods	245
	Bibliography	269
	Index	271

1

Introduction to Agro-Processing

Agro-processing is regarded as the sunrise sector of the Indian economy in view of its large potential for growth and likely socio economic impact specifically on employment and income generation. Some estimates suggest that in developed countries, up to 14 per cent of the total work force is engaged in agro-processing sector directly or indirectly. However, in India, only about 3 per cent of the work force finds employment in this sector revealing its underdeveloped state and vast untapped potential for employment. Properly developed, agro-processing sector can make India a major player at the global level for marketing and supply of processed food, feed and a wide range of other plant and animal products.

By the middle of the nineteenth century, common agro processing industries included hand pounding units for rice, water power driven flour mills, bullock driven oil ghanies, bullock operated sugarcane crushers, paper making units, spinning wheels and handloom units for weaving. In British India, during the year 1863, a note was written by the Governor of Madras state, Sir William Denison to the government of Madras state for laying greater stress on agriculture and agro processing. Based on this, a set of improved machinery was brought from England for demonstration and adoption. It included threshing machines, winnowers, chaff cutters, besides steam ploughs, steam harrows, cultivators, seed drills and horse hoes.

The demonstration continued at Saidapet near Madras till 1871 with little outcome. Importance of agro-processing sector was first realised and documented after the disastrous famine of Bengal during 1870's. Report of the Famine Commission, set up by the British Government, in its report submitted in 1880, clearly stated the need for agricultural improvement and improved post harvest infrastructural development specifically, rail network. Need was also felt for incorporating chemical interventions in the agricultural sector and precision farming through agricultural mechanisation manned by engineers.

The Imperial Agricultural Research Institute, Pusa; Indian Veterinary Research Institute, Mukteshwar; Dairy Research Institute at Bangalore; Poona Agriculture College; Public Agriculture College, Saidapet (Madras); Sibpur Engineering College (Bengal) etc. Horticultural Research Station was created at Chaubatia (U.P.) in Kumaon Hills for horticultural research including packaging and transportation improvements. Post independence era in India witnessed rapid growth in agro processing sector specifically during 1980s. It followed the first phase of the Green Revolution that had resulted in increased agricultural production and the need for its post harvest management.

The importance of the sector was realised by the business community leading to diversification from grain trading to processing. Lead was given by the rice processing industry, followed closely by wheat milling, paper and pulp industry, milk processing sector, jute industry, sugarcane processing and oils extraction through solvent plants.

In some areas like the solvent extraction industry, the growth in installed processing capacity has been far higher than the supply of the raw materials. However, in other areas like fruits and vegetable processing, the growth has not been encouraging on account of poor demand for processed products by the consumers. In such cases, the industry has also not been able to develop the demand adequately.

Agricultural Production Trends

At the start of the twentieth century, Indian agriculture was in a stage of subsistence. By the year 1925-26, the total area under some major crops in undivided British India was: rice - 32 mha, wheat - 9.6 mha, sorghum - 8.2 mha (Royal Commission on Agriculture). The yields were very low. In the year 1950-51, India produced only 50 million tonnes of food grain and a variety of other crops.

By the year 2000-2001, India started producing about 700 million tonnes (Mt) of biological materials per year including food grains, oilseeds, fruits, vegetables, sugarcane, milk, eggs, meat, fish, tea, coffee, fiber crops, floricultural produce, forest produce and so on. The country has diverse agro-climatic conditions and consumer preferences and hence it produces a vast variety of agricultural and livestock materials.

Post Harvest Management

On account of poor post harvest management, the losses in farm produce in India have been assessed to be of a very high order. Various studies have estimated post production losses in food commodities to the tune of Rs. 75,000-1,00,000 crore per annum. It is also estimated that the extent of losses could be brought down to less than 50 per cent of the existing level on proper transfer and adoption of agro processing technology. For reducing the rest of the losses, new initiatives need to be called for. Hence, it would be in the long term interest of the economy to invest in developing suitable infrastructure such as proper grain storage structures, cold stores and processing systems to avoid the losses.

R&D Organisations

Significant increase has taken place after early fifties in the number of institutions engaged in agro processing research. In the area of teaching, presently there are 18 universities/ colleges offering first degree, 11 offer post graduate and 7 offer Ph.D. degree.

Among R&D organisations in the area of agroprocessing, ICAR has 17 Institutes with some component of Post Harvest Technology (PHT), CSIR has 3 laboratories; State Agricultural Universities have 18 programmes, IITs have 2 programmes and 11 other organisations have similar programme. Some of the leading government funded R&D Institutes (based on their infrastructure and sanctioned scientific manpower) in 2000-2001 were: CFTRI, Mysore; CIPHET, Ludhiana; IARI, New Delhi; NDRI, Karnal; DFRL, Mysore; CIAE, Bhopal; IIT, Kharagpur; GPBUA&T, Pantnagar; IGMRI, Hapur; TNAU, Coimbatore; PAU, Ludhiana; GAU, Anand; RAU, Udaipur; BCKV, Kalyani; OTRI, Anantpur; PPRC, Thanjavur; MERADO, Ludhiana; MPKV, Rahuri; ILRI, Ranchi; IVRI, Izatnagar; NIRJAFT, Kolkata; CIRCOT, Mumbai; IISR, Lucknow; IGFRI, Jhansi; KVIC Mumbai; HBTI, Kanpur and PHT Institute, Pune.

The ICAR has a system of All India Coordinated Research Projects (AICRP) in various important areas. In the field of PHT, there are 4 AICRPs: (1) All India Coordinated Research Project on Post Harvest Technology (21 centers in the country, coordinated from CIPHET, Ludhiana), (2) Processing, Handling, and Storage of Jaggery and Khandsari (5 centers, coordinated from IISR, Lucknow), (3) Application of Plastics in Agriculture, Plant Environment Control & Agricultural Processing (5 centers, coordinated from CIPHET, Ludhiana) and (4) Post Harvest Technology of Horticultural Crops (8 centers, coordinated from IARI, New Delhi).

Also, there are other AICRPs that have a component of PHT. These are (1) Renewable Sources of Energy for Agriculture and Agro-based Industries (16 centers, coordinated from CIAE, Bhopal), (2) Farm Implements and Machinery (21 centres, coordinated from CIAE, Bhopal), (3) Utilisation of Animal Energy with Enhanced System Efficiency, (4) Human Engineering and Safety in Agriculture. Besides these, there are about 60 ad-hoc research projects operating in different SAUs, universities, IITs, CSIR Institutes and other laboratories that have been working on problems related to PHT.

A number of Universities have programmes in the area of agro processing. Some of the state governments also have been supporting R&D activities on agro processing in a number of their laboratories/departments. Although nearly 2000 scientists were associated with agro-processing R&D in the year 2000-2001, only about 200 out of them could be considered as full time R&D workers in agro-processing.

R&D Work in agro-processing carried out in India during the last 50 years categorised as follows:

- Studies on physical, biochemical, nutritional, and engineering properties/characteristics of different food, feed, fibre, and industrial raw materials.
- Response studies of different biological materials w.r.t. their storage, handling, and moisture conditioning.
- Refinement of traditional equipment and processes for production of different foods, feeds, fibres and fuel materials for better quality, higher capacity, energy efficiency, and reduced drudgery to workers.
- Development of new produces and processes for better nutrition, convenience and taste.
- Enhancement of shelf life of the produces, safe storage/ packaging and development of better performing materials.
- Better economic utilisation of agricultural residues, by-products and recycling of wastes.
- Design and Development of instruments and equipment for post harvest operations and their evaluation, feasibility analysis, field trails/ multilocation evaluation etc.
- Design, layout planning and development of pilot plants, agricultural produce bulk handling systems and area specific agro-processing models.
- Studies and modeling/simulation of post harvest systems and industry for the purpose of optimisation, forecasting and policy analysis.

— Energy auditing and use of non-renewable sources of energy for post harvest operations.

— Product quality analysis, sensory evaluation and consumer acceptance studies.

— Work conditions, safety and pollution control.

Among large number of technologies developed, the most popular ones include:

— Agriculture produce refinement equipment such as, cleaners, graders and driers for on-farm operations as well as industrial operations.

— Processes and equipment for parboiling of rice, preparation of puffed rice and flaked rice.

— Development of processes and equipment for processing of pulses to produce dhal for higher recovery and better quality.

— Development of driers using agricultural residues, by-products and solar energy.

— Adoption and development of processes, and equipment for production of protein rich produces such as full fat soy flour, soy drink/ soy milk, soy paneer (TOFU) and soy fortified baked products.

— Development of equipment such as, leaf cup and dona making machine, multipurpose mills, mini flour mill, grain pearlers, maize dehuskers, shellers, groundnut decorticators, fruit graders, juice extractors, high recovery mechanical oil expellers and improved storage structures for cereals, pulses, oilseeds, onion and potato.

— Processes and equipment for production of high quality ground spices and spice mix, development of raw materials and processes for production of instant sweets, curries, snack foods, instant soft drinks, idli, dosa, sambhar mixes/powders, egg powder, production and packaging of milk products such as shrikhand, butter milk, paneer, ghee and sweets.

— Equipment for high recovery of sugarcane juice, processes for production of high quality jaggery and liquid jaggery.

— Processes, equipment and pilot plants for production of various industrial raw material from lac including dyes and pharmaceutical products.

— Improved technology for processing of jute sticks to yield jute fibre and impregnation, preparation of jute based textile materials and bags.

— Control of stored grain insects by using chemical and physical methods, storage structures for onfarm, trade, and process plant level operations.

— Processing and canning of meat, meat products and fish.

Some work has also been done in the area of processing forest produce such as oil extraction from oil bearing materials, collection and processing of resins and production of dyes, chemicals and pharmaceutical products. The latest developments have been in the area of floriculture. Due to high export potential, R&D work has been initiated at some centres on pre-cooling, packaging, and transport of cut flowers and low cost designs of green houses. Agro-processing models have also been developed for some of the agro-climatic regions in the Country.

In the area of agro-processing of fruits and vegetables, development of tools and techniques for harvesting, precooling of freshly harvested produce, minimal processing, controlled ripening, juice extraction, concentration and storage has been done. Similarly, in the area of spices & condiments, floriculture, production of mushrooms, honey, eggs and fish, technologies have been developed for post harvest loss reduction and value addition.

Agro Processing Sector

Starting with a small number of processing facilities in 1950-51, a fairly well spread network of processing facilities has developed in the Country. Various estimates suggest the number of processing units in 2000-2001 as: atta chakkis and small hammer mills - 2,70,000, rice hullers - 90,000, rice shellers - 11,000, huller-cumshellers - 12,000, modern rice mills - 30,000, bullock/ electricity operated oil ghannis - 2,00,000, oil expellers - 55,000, dhal mills - 12,000, roller flour mills - 700, rice flaking and puffing units - 2,000, bakery units - 54,000, solvent extraction plants - 700, vanaspati plants - 100, fruits and vegetables processing plants - 5,000, dairy plants - 450, cold storage units - 3,000, licensed units in organised sector for meat processing - 165, pork processing units - 144, fish processing units - 18 and so on.

Major problems faced by these units have been:

(a) low capacity utilisation,

(b) poor recovery of the finished product from the raw materials,

(c) problems of arranging adequate working capital and its management,

(d) low product quality and

(e) unreliable assured power supply.

Strong R&D support will have to be continued to over come these and many other problems to ensure that our agro-processing technology becomes competitive at the global level. As stated earlier, inspite of the problems, agro-processing technology in India has continued to make steady progress towards modernisation.

Commodity Wise Status of Agro-Processing Industries

The commodity-wise growth of agro-processing industries in the country during the years 1950 to 2000 has been as given below.

Rice Processing Sector

Starting with 20.6 Mt of rice production during 1950-51, the country has come a long way to produce about 89.48 Mt of rice in the year 1999-2000. Similarly, in processing sector, the technology has undergone significant changes. Earlier, hand pounding, pedal operated system and Engleberg huller units were common for milling of paddy. By the year 1998-99, there were nearly 30,000 modern rice mills using rubber rolls for paddy dehusking. Of these, more than 5,000 are large rice mills with parboiling facility and nearly 100 have colour sorters for removal of discoloured rice for export market.

Innovations in rice processing include improved process of parboiling developed at IIT, Kharagpur; CFTRI, Mysore; PPRC, Thanjavur and other R&D centres. Starting from sun drying, the technology for drying of paddy now includes use of a variety of driers, specifically for parboiled paddy. Continuous flow LSU type driers have been most commonly used units followed by tray driers (batch type). Thermic fluids are used as medium of heat transfer for heating the air used for drying in a large number of rice mills. Though efforts have been made to improve the rice hullers, limited success was achieved in improving their performance with respect to reduction in broken percentage.

Rubber roll technology for dehusking has now been well established. Efforts are ongoing to find use of tafflon to replace rubber rolls for economy. Several types of rice bran stabiliser have been designed and tested. Chemical method developed at CFTRI, Mysore; steam heating at IIT, Kharagpur, electrical heating method developed at Pantnagar could find limited applications in Industry. Stabilisation through extrusion technology has also been tried with limited application of expanders. Among most common value-added products of rice include puffed and flaked rice used as snack foods.

Rice and wheat form the major part of government operated procurement system and storage. In the month of March 2001, the total stocks of rice and wheat in FCI/ CWC and other government owned godowns were about 35 million tonnes for the public distribution system, for processing industry and for future use. Significant achievements have been recorded in packaging technology for milled rice for ready-to-cook applications in domestic market and export. Quick cooking rice has been developed at DFRL, Mysore and CFTRI, Mysore. The technology is being used for making available food supplies to defence personnel in boarder areas under war or war like situations.

Rice is partially cooked and packed under highly sanitary conditions. It is autoclaved and supplied for safe use upto 6 months of period. Rice bran oil is a common form of edible oil besides its application in industry specifically as soap stock. For utilisation of rice husk, a number of efforts were made at GBPUA&T, Pantnagar; PAU, Ludhiana; CFTRI, Mysore; IIT, Kharagpur; TNAU, Coimbatore; PPRC, Thanjavur and other R&D laboratories. Its application as sources of furfural, high grade silicon, insulation material, particle board and as source of fuel have been well demonstrated. However, rice husk is being used only as source of fuel in rice mills, in making particle boards, in poultry houses as bedding material and in land fills.

Similarly, paddy straw has found limited applications as cushioning material in packaging of fruits and for preparation of soft boards. It is extensively used as cattle feed in many parts of the country. The upcoming areas in rice processing R&D include high capacity dehuskers and more efficient polishers improved technology for storage of paddy and rice, onfarm/ community level drying of paddy, mechanical handling systems for grain markets and millers, cold storage of rice and down stream products, products diversification in the form of flakes, puffed rice, snacks, bakery items, quick cooking and ready-to-eat rice etc.

The recovery of brown rice as obtained from the hullers, shellers, and modern mills could be in the range of 62-64, 65-67 and 68-70 per cent, respectively. The potential yield of rice is 70-72%. The need is therefore, to promote modern rice mills and develop milling technology for fine rice. If all paddy is milled in modern rice mills, 3 million tones of additional rice worth Rs. 15,000 million could be obtained.

Wheat Processing Industry

Wheat is a major crop of India. In the year 1950-51, the country produced 6.5 Mt of wheat, that has increased to 76 Mt by the year 2000-2001. India has emerged as the second largest producer of this cereal in the world. Wheat contains 12% bran, 3% embryo and 85% flour. It is mainly processed for flour (atta), maida, suji and dalia. In last 50 years, harvest and post harvest technology of wheat has advanced substantially.

The most significant development has been the use of self propelled harvester combines used for harvesting and threshing of wheat. From a small figure of about 20-30 combines during 1950-51 imported from USSR by the State Farm Corporation of India, the number has now grown to nearly 6,000 combines. In the year 1998-99, there were about 27 lakh atta chakkis (7.5-10 kW rating) and 700 roller flour mills in the country. This number has risen from 53,000 atta chakkis and 200 roller flour mills in 1971-72. The figures were much lower 50 years back.

The industry could grow on account of R&D inputs starting from the design and development of a variety of threshing machines. Mud bins, wooden plank and mud plastered bins, gunny bags and metal bins have been in use by the farmers for storage of wheat for food and for seed purposes. The traders and government agencies use gunny bags and godown type structures for storage of wheat. For transit level storage, CAP structures have been in use. Metal bins have gained popularity among farmers in the capacity range of 0.2-1.0 tonne of grain storage. As wheat is usually harvested at low moisture content, drying has not been a major problem except for untimely rains.

A number of commercial organisations have been offering processing units for handling, cleaning, grading, drying, storage, treatment and bagging of wheat for seed and food applications. Wheat is now increasingly being used in the form of bread, biscuits, suji and atta. Wheat flakes and puffed wheat as breakfast cereals has been gradually picking up. In the area of wheat milling, Central Food Technology Research Institute, Mysore; Central Institute of Agricultural Engineering, Bhopal and a number of other R&D institutions have developed mini flour mills for higher efficiency in small scale.

Traditionally used smaller size atta chakkis may face problems of declining clientele. Better mechanised chakkis (with lower pollution level and better energy efficiency) are likely to increase in number. The number of roller flour mills is also likely to increase steadily, however, majority of the mills may continue facing the problems of low capacity utilisation and working capital constraints. These units would need to function through vertical integration of operations for sustaining profitability and achieve cost reduction through appropriate automation and computerisation.

Increase in demand is also expected in grain handling machinery, silo systems in grain markets and seed processing machinery. Trends in consumer preferences suggest increasing demand for baked products. Demand for bread is likely to grow faster than the demand for biscuits. Presently bread is consumed mostly in large cities. Its consumption is expected to grow in smaller towns also. States with higher per capita income would continue to lead in the consumption of baked products. Among diversified products, full bran wheat bread has also been gaining popularity.

Production of Coarse Cereals

Production of coarse cereals has risen from 15.4 to 32.0 Mt between 1950-51 and 2000-2001. The growth has not been as rapid as in case of wheat and rice. It is because of low profitability of these crops for farmers. Till 1950s, we were dependent on manual methods of harvesting of these crops, bullock treading, storage in mud bins and gunny bags and milling by manual chakkis or water mills. By the year 1998-99, power operated equipment were available for all operations including threshing, pearling and milling.

For storage of coarse cereals, metal bins have been designed at IGSI, Hapur; CIAE, Bhopal; CFTRI, Mysore; PAU, Ludhiana and several of the other R&D Centres. For drying of freshly harvested HYV sorghum or maize, hot air driers using agriculture residue as a source of fuel are now in use. Technology has also been developed for production of value-added products from coarse cereals such as extruded snacks developed from ragi at CFTRI, Mysore; ragi based snacks at UAS, Bangalore and IIT, Kharagpur; corn products at GNDU, Amritsar; ready to-eat traditional foods with storage life of 6-9 months at DFRL, Mysore and sorghum-soybean fortified foods at IIT Kharagpur.

The trends indicate that coarse cereals are now increasingly used as cattle feed, speciality/ occasional foods, and industrial products such as starches. Efforts are required to develop high yielding varieties of coarse with desired characteristics for different uses and to explore new food uses. Safe storage of the flour produced from most of the coarse cereals has been a problem due to its high degree of perishability. This problem needs to be solved.

Milling of Pulses

India produced 8.4 Mt of pulses in the year 1950-51. The production grew to a level of about 14 Mt by the year 2000-2001. Starting with nearly 500 dhal mills in the country in 1950-51, there were about 15,000 dhal mills of 100-500 TPD capacity in the year 2000-2001. Pulses were generally stored in gunny bags or in small tin containers under straw cover during 1950s. By the year 2000-2001, metal bins and gunny bags (with profilectic treatment by insecticides) were in use. Research at CIAE, Bhopal; CFTRI, Mysore; JNKVV, Jabalpur and GBPUA&T, Pantnagar has revealed that pulse grains need to be stored at 20-22 degree Celsius in partially airtight containers at 8-10 per cent moisture content for long duration storage.

A number of plant based mild insecticides and insect repellents (such as, neem seed powder) have been developed for safe storage of seeds. In the area of milling of pulses, CFTRI developed a dhal mill that has the advantage of not being dependent on natural sun shine. It involves subjecting the pulse grain to high temperature (120 degree Celsius) for short time and the dehusking by carborundum rollers resulting in higher dhal recovery. For small entrepreneurs in rural areas, dhal mills have been designed at CIAE, Bhopal; PDKV, Akola; IIPR, Kanpur, TNAU, Coimbatore; GBPUA&T, Pantnagar and

CFTRI, Mysore. These units in specific regions have gained popularity as these are low investment machines which can be owned and operated with low risk.

In a number of dhal mills, improved machinery including cleaners, graders, magnetic separators, washers, driers, polishers, colour sorters and packaging systems are being used. With complete phasing out of hand operated dhal chakkis, commonly used during 1950s, the technology has turned fully mechanised and more-and-more urban based. There is a need to evolve more efficient machines and processes for pre-treatment of the grain, dehusking, sorting, polishing and packaging in order to improve dhal recovery and consume less energy. Also, there is a need for product diversification and development of technology for quick cooking and ready-to-eat dal.

Oilseeds Processing

Besides, animal based fat specifically obtained from milk and milk products, edible plant oils have been the major source of oils and fats for most of the population in the country. In the year 1950-51, the country produced 5.2 Mt of oilseeds. Production by the year 2000-2001 had increased to 24.5 Mt. In the year 1950-51, most of the oilseeds were crushed in either bullock operated oil ghanies or a few mechanical oil expellers. Both of these resulted in high volume of edible oil left in the cake. By the year 2000-2001, there were nearly 2.5 lakh oil ghanies, 60,000 oil expellers and 700 solvent extraction plants.

Besides, there were 200 oil refining units in the country and 100 units for production of hydrogenated oil (Vanaspati). Per capita availability of edible oils is still very low at 8.0 kg per capita per year in the country. Out of this, 2 kg/capita is imported oil. R&D Institutions in the country have been working on pre-treatment of oilseeds for higher recovery of oil. Steaming has been found as one of the most useful methods for pre-treatment. On mechanical oil expellers, a number of Institutions including CFTRI, Mysore; CIAE, Bhopal; RAU, Udaipur; OTRI, Anantpur; HBTI, Kanpur and KVIC, Mumbai have done significant work on mechanical oil expellers. KVIC tried to improve the design of bullock operated oil ghanies to make them suitable for operating on 1.5 kW electric motor.

Hence, the capacity of these units has improved significantly. These units produce pungent oil that is being traditionally liked by the consumers. The oil produced from ghanies is also being mixed with oil obtained from mechanical oil expellers to produce pungent oil. In the area of solvent extraction of oil from traditional and nontraditional oilseeds, a large number of researches have been done. A number of chemical solvents have been tried. However, for reasons of economy, food grade hexane has been accepted commercially for solvent extraction of edible oil.

Work done at different R&D Institutions also reveals that for long duration storage, oilseeds need to be put in metal containers with limited aeration. Hence, metal bins designed at IGSI, Hapur; CIAE, Bhopal; GBPUA&T, Pantnagar and other centres have become popular. Due to shortage of edible oil in the country, efforts have also been directed to obtain edible oil from non-traditional sources including rice bran and oil palm. On refining, the quality of these oils has been reported satisfactory for edible purposes. In the area of packaging of edible oils, significantly rapid growth has been recorded specifically in commercial sector. Polypacks and plastic containers have gained popularity over traditionally used metal containers about 30-35 years ago.

The future areas of research include application of bio-technology for enhancing yield of edible oil from different oilseeds, application of de-oiled cake for food purposes through protein isolation and health applications of edible oil for treating various physiological disorders. Production of oilseeds is 24.5 million tonnes. Out of the total production, 7% is used for seed, 8% for food, and 85% for oil extraction. Export of meal/oilseeds cake has been worth Rs. 15,000 million. Refinement of meal/ cake for food products development could be of high importance.

Oil expeller with lighter weight, high energy efficiency and capable of extraction up to 90% oil and above needs to be developed for decentralised oil milling. Hydraulic press, batch solvent extraction, extrusion-expelling and physical refining, also need to be considered and tried. Besides other oilseeds, soybean has gradually become an important crop of India. Its production is around 5.3 million tonnes. Soybean is a special legume. It has 40% protein and 20% oil. India has 154 solvent extraction plants and 60 soyfood units.

Average recovery is 17.7% for oil and 82.4% for meal. Soymeal contains about 48% protein. Its export has been worth Rs.15,000 million/ year. Soy foods are nutritious and economical and must be promoted. A strategic plan for expanded and diversified use of soybean for food and feed in India for the next 25 years should be made and implemented. This crop has a great potential to enhance nutrition and health of the people and alleviate poverty.

Processing of Fruits and Vegetables

Joint effort of R&D institutions, farmers, government agencies and the trade has resulted in India emerging as a major producer of fruits and vegetables in the world. In the year 2000-2001, the country produced about 45 millions tonnes of fruits and 80 millions tonnes of vegetables. It was next to China in production of vegetables and topped in production of fruits. However, the growth in post harvest sector has not kept pace with the production.

Even during the year 2000-2001, there were only 6,000 fruits and vegetable units in the country that had grown from a figure of about 1,000 during 1950-51. Less than one per cent of the total produce was processed, though the installed capacity of the processing industry has grown steadily from 0.27 Mt in 1980 to about 3 Mt in 2000-2001. Significant developments in technology include better understanding of the process of ripening of fruits, optimum harvesting time, pre-cooling of freshly harvested produce, cold storing of the raw fruits and vegetables, sorting, cleaning, waxing, packaging technology for fruits.

At CFTRI, DFRL, IIHR, Bangalore; IARI, New Delhi; GBPUA&T, Pantnagar; IIVR, Varanasi and HPKV, Palampur;

a number of technologies have been developed. Most significant work has been recorded in the technology for ripening of the fruits under controlled conditions. Production of juices and value-added products including jams, jellies, pickles, canned products etc. has become a commercial success. The industry using indigenous technology includes units engaged in juice extraction, concentration of juices, canning and production of several of the products like jams, jellies, canned fruits, dried vegetables etc.

Technology is still being imported for establishment of large scale exported oriented units for production of items like banana paste, concentrates of various fruit juices, sorting, cleaning, washing, waxing and packaging of raw fruits and vegetables. By the year 1998-99, share of different products in the total processed fruits and vegetables was; pulp and juice 27%, jams and jellies 10%, pickles 12%, ready to-serve beverages 13%, syrups 8%, squashes 4%, tomato products 4%, by canned vegetables 4% and other products 18%.

The industry has been facing problems of low capacity utilisation, technological obsolescence and marketing. It has to work under the constraints of high fluctuations in raw material quality and fluctuating market price, poor technology for handling and storage, inadequate R&D support for product development, high cost of energy and uncertainty in availability of adequate quantity for processing purposes, inadequate and expensive cold chain facilities and varying requirement of processing conditions from one material to another. Future R&D has to focus on the issues of economically producing value-added products and product diversification, besides the issues mentioned above.

Processing of Sugarcane

Sugarcane production was 310 Mt in the year 2000-2001. About 80% of the cane produced is milled, about half for the production of refined white sugar in the organised sector with the sugar mills located in the production catchments in public, private and cooperative sectors and about 42% for the production of Jaggery and Khandsari. Based on sugar recovery, minimum price scheme has been introduced. Mills have loose tie-up with the growers, some of them provide critical input support to the growers. Apparently, it is working well. But there have been cases where farmers burnt their crops in the absence of remunerative prices. For Jaggery, canes are crushed, clarified and concentrated.

Gur as sweetener has better nutritional profile than white sugar. It is possible to refine the process and the product for greater competitiveness and realise export potential specially where people of Indian origin are located. Energy efficient furnaces, concentration pans, clarificants, moulds and storage are needed for Gur. Khandsari units used open pan in place of vacuum pans for concentration and the sugar obtained is of lower quality compared to white sugar from mills. Sugar recovery in Khandsari is much lower. These units depend on grid supply or diesel generators for mechanical/electrical power or both when grid power or both when grid power supply is erratic and diesel gensets are kept as standby power sources.

This increases the cost of production of Khandsari. Bagasse, tops, dry leaves and molasses are by-products. Modern sugar mills with co-generation meet their entire energy needs, both thermal and electro mechanical from these bagasse fired boilers—steam turbine units. They feed extra power to grid or save 15-20% bagasse for the use as feedstock or paper making. Jaggery promotional and regulatory measures have been taken by the Government to improve quality and production. Large number of sugar mills are using outdated processes and equipment, some of them not only use entire bagasse but also use wood.

Processing of Cotton

Cotton is a natural textile fibre. Traditional cotton textile industry could not face onslaught of modern high speed spinning, weaving and surface finish technologies. Small scale textile industry supported by Swadeshi and Khadhi and Village Industries Commission face serious labour problems also. Cotton seeds are valued as feed and oilseed and the stalks are used as fuel. However, stalks yield excellent paper and pulp, particle boards and microcrystalline cellulose (MCC). Cotton hulls also yield good particle board and furfural. Cotton willow dust can be used for production of bio-gas. Cotton wastes can be used for mushroom production. There is scope for income and employment generation if cotton stalks are utilised for pulp and paper making.

Processing of Jute

Jute has the distinction of having ushered India into industrialisation era. Both jute production and manufacture of jute-based products are highly labour intensive, concentrated mostly in Eastern India. Mini jute carding and spinning mills have now been developed which allow decentralised production of utility items from jute but these are not popular yet.

For each tonne of jute, 2-3 tonnes of jute sticks are produced. Chemically these resemble hardwood. Sticks are traditionally used as fuel wood and low cost structural material. Jute sticks yield excellent particle boards and the technologies are now fully commercial. Jute sticks are a good feedstock for paper pulp. The sticks can also be used as fuel for steam and power generation.

Processing of Animal Produce

Meat and poultry production in India has been about 4.6 Mt per year with goats and sheep contributing 54%, buffalo and cattle 26%, poultry 13% and pig 7%. It is mostly used fresh. Efforts are on to develop infrastructure for export of both fresh and processed meat and poultry. Production is essentially decentralised and rural based.

Poultry has done well remaining in rural sector and developing network of marketing in distant remunerative markets. Hygiene in slaughter houses and use of blood, viscera and other wastes is not satisfactory. The meat from culled birds, goats and buffaloes is tough textured, better suited for processed meat products. However, there is no tradition of using processed meat products in India, yet.

Fish and Fish Products Processing

India, with its 7,500 km long coastline and an exclusive economic zone of 2.02 million square km; 191,024 km of rivers & canals and 4.4 million hectares of reservoirs and fresh water lakes has an enormous potential for fisheries. In 1999, the country had an estimated 1,81,284 traditional fishing crafts; 44,578 motorised traditional crafts, 53,684 mechanised fishing boats and about 200 deep-sea vessels in operation.

Fish processing in India is done almost entirely for export. Open sun dried fish and fish meal are the only major exceptions. At present India has freezing units, cold stores, ice plants, canning units and fish meal plants. Capacity of most of these processing and storage units in small when compared to the facilities in fish processing industry in technologically advanced countries.

The total fish processing and storage facility in India grossly is inadequate compared to the potential for fish production and processing. Inland fisheries need low cost palletised feeds and special containers to transport fingerlings and fish. More rearing ponds are needed. Techniques to reduce seepage loss of water have to be introduced. Obsolete fishing gear needs replacement with better gear. Extensive network of refrigerated handling, transport, storage and retailing has to be put in place. Also, we have to make better use of fish waste and by-products.

Commercial Crops

The commercial crops include spices, condiments and crops such as gorgon nut (makhana), water chestnut, bettle leaves, tobacco etc. Post harvest operations of these crops are highly energy intensive and there is a scope for reducing energy consumption and improvement of quality through proper cleaning, grading, drying/dehydration, milling, grinding and other operations. India has been a leading producer, consumer and exporter of spices like black pepper, cardamom, chilies, spice oils and oleoresins. It produces about 3.0 Mt of spices valued at over Rs. 60,000 million.

About 7% of the total production is exported. Contribution of R&D to PHT of spices includes equipment and processes for cleaning, grading and packaging of whole spices and production of value-added products such as oleoresins and spice oils. Institutions like CFTRI, DFRL, Indian Institute of Spices Research and some of the SAUs including TNAU, Coimbatore have contributed significantly to this development.

Projected world trade in spices in 2001 AD was estimated to about 6.25 lakh tonnes, valued at US$ 3 billion and projected export from India at that time was about 10% of the world trade. To achieve and maintain India's share in the trade, the quality of spices and their products will have to be improved. New products like dehydrated pepper, freeze dried green pepper, ginger candy, ginger beer/in-brine/ squash, ginger flakes have to be developed. Development of internationally accepted quality products, packed under hygienic conditions need attention in this context. Similarly, in the area of PHT of other crops, contribution of R&D basically has been on raw materials refinement, product quality enhancement and diversification.

Plantation Crops

Plantation crops contribute substantially to the national economy with export earning of Rs. 12.4 billion. Coconut alone contributed Rs. 1.72 billion by way of exports during 1996-97. However, the coconut based industry in India has been in the infancy stage. There is considerable scope of product diversification, viz. production of coconut milk and milk powder, coconut cream, shell powder, shell charcoal etc. Coconut wood utilisation needs more attention. In case of other crops, financially viable technologies for product diversification need to be developed. Such products are arecanut fat, tannin, areuline, other chemicals from arecanut, honey/ chocolate coated or salted kernels from cashewnuts and value added products from by-products.

The post harvest operations in these crops need to be mechanised. Though the technology has been developed for desiccated coconut, coconut cream and other products, it needs refinement. At CPCRI, Kasargod, a coconut dehusker has been developed for manually opening the nuts. Another motorised unit is under development. Copra drier using LDPE cover and batch type hot air copra drier using agricultural waste as source of fuel have also been developed at CPCRI, Kasargod; KAU, Thrissur and TNAU, Coimbatore. In case of the plantation crops like oil palm, necessary efforts are required for processing and value addition, especially with regard to quality of products, energy inputs, packaging etc. to meet the international quality standards and to reduce the cost of production. Processing of cocoa beans at small scale also needs attention.

Processing of Medicinal and Aromatic Products

The plant based pharmaceuticals, herbal medicines, perfumery, cosmetics, fragrances and food flavour industries have recorded a phenomenal expansion in last 50 years and as a result, this sector figures in high annual growth rate industries in agri-business. The market for plant based pharmaceuticals in the year 1994 was estimated to range between US$ 32-43 billion.

The world essential oil production at raw materials level was estimated to be about Rs. 32 billion of which 55-60% goes to food flavours, 15-20% as fragrances and the remaining is broadly used as starting raw material for isolation of aromatic chemicals. In terms of market share in production value, India is sliding downwards and presently stands at sixth rank with only 6% share in world trade. The R&D work on PHT of medicinal and aromatic plants had been confined to the IBRI, Lucknow and a few of the CSIR and ICAR laboratories.

The thrust has been harvesting of the plants, curing/ drying, and extraction of the medicinal and aromatic substances. The export earnings could be increased by innovations in the field of post harvest technology for increasing productivity and improving quality. In case of medicinal plants, studies need to be conducted to develop testing procedures/analytical facilities to meet stringent international standards and to carry out product/process development for low cost chemicals from both raw materials and other by-products.

Apiculture Produce Processing

Bee-keeping i.e., rearing the bees in artificial hives to produce honey and other products offers an immense potential for providing employment to rural folk in India where many evergreen and moist deciduous forests, orchards etc. constitute good bee keeping areas. The unique feature of bee keeping is that the capital investment required is small and unlike many other industries, it does not need raw material in usual sense as nature offers the same in the form of nectar and pollen. The equipment required, viz; bee boxes of standard sizes, honey extractor smoker, hive tools etc. have been researched and improved in design and these can be manufactured even in small rural carpentry and black smithy shops. Improved bee hives have been developed which make honey production much easier than the traditional long hanging hives.

In general, equipment like smoker, comb foundation sheet machine, honey extractor, queen excluder, honey tank and uncapping equipment have been developed by R&D organisations namely, KVIC, Mumbai; PAU, Ludhiana and IARI, New Delhi. There is need for R&D to develop suitable equipment in this reference and for product diversification. Good work has also been reported by GNDU, Amritsar in improving quality of honey through proper processing and measuring its bio-chemical and engineering properties that could be used by processing industry.

Traditional Food Processing

India has a very strong base of traditional food products, which have been developed under varied agroclimatic, geographical and socio-cultural situations over the centuries. Besides, conventional chapaties, these may include expended, puffed, flaked, extruded, fermented products, sweets, instant mixes, breakfast foods, bakery products, beverages, health and special foods. The production of traditional foods during 1996-97 has been estimated nearly 30 times more than that of all western style high cost processed foods in the Indian market.

There is an urgent need to upgrade the conventional foods technology so that the industrial manufacturing of products can be promoted and the scope of marketing expanded. There can be substantial domestic and export demand for traditional foods. Production of Bikaneri Bhujia for export and frozen Idli for domestic markets are some of the successful examples. A chapatti making machine developed at CFTRI, Mysore for defence canteens and hotels is a fine example of modernisation of traditional food sector.

Similarly, long life chapatti and parotha technology developed at DFRL, Mysore has been a success story for providing food of liking for many defence personnel working in remote and frontier areas. Technologies during last 50 years have also been developed for gulabjamun mix, idli and dosa mixes and a variety of other food items to suit to Indian palate. R&D has contributed significantly in rapid growth of processing units and trade in traditional food sector however much more needs to be done.

Floriculture

Flowers and plants have always been an integral part of human living. Besides their aesthetic importance, they are also useful in improving the quality of life. Ornamental plants play a very important role in environmental planning of urban and rural areas for abatement of pollution, social and rural forestry, wasteland development, aforestation and landscaping of outdoor and indoor spaces. Floriculture is also an important agri-business with potential for export trade.

The area under floriculture in India has increased to nearly 40,000 ha, which constitutes around 17% of total global acreage. Inspite of such a large area, production value is very low. The quality of Indian produce is poor and not acceptable in international market. The produce quality deteriorates further due to improper packaging, storage and transportation. Major contribution of R&D in this area has been in the form of raising varieties that are more attractive and flower life is longer. Also, technology has been developed at IIHR, Bangalore; IARI, New Delhi; UAS, Bangalore; HPKV, Palampur; GBPUA&T, Pantnagar and PAU, Ludhiana for longer shelf life of cut flowers.

Floriculture is largely an export oriented agroindustry. There are 14 flowers in the world cut flower trade. The trade is growing at the rate of 15% per annum. Yet Indian exports are limited only to a few flowers namely, Gladiolus, Chrysanthemums, Jasmine and Orchids. India's share in the world floriculture trade is a minuscule with 0.59% exports during 1992-93 valued at Rs. 149.1 million. Cultivation of high quality varieties under protected conditions, proper tools and equipment, appropriate packaging and storage can create a niche for Indian flowers in the world market.

Trends and Opportunities of Agricultural Exports

India has been a traditional exporter of raw agricultural products like spices. Export of raw products has resulted in huge loss to Indian economy. After GATT agreement and WTO membership, processed products manufactured as per international norms only offered at competitive prices, can be exported. However, our processed products mostly do not meet the international standards. India's share in over US$ 300 billion world trade in agricultural commodities is less than 1%.

Agricultural exports used to be of the order of 30.6% of the total exports during 1980-81, which came down to 19.4% by 1990-91. Currently, it is at about 16% due to rapid growth in other sectors as well. Processed fruit and vegetable products have considerable export potentials and if it is properly utilised, growers, processors, traders as well as national economy will benefit. It requires correct assessment of world market, high quality of raw produce, high quality of processed product and competitive production cost.

Quality Control

Food processing industries cover a large spectrum of products of plant and animal origin. Quality has got to be maintained for domestic as well as export markets. In this respect, a number of organisations have come up for the formulation of standards and for monitoring their quality. These can be classified into two groups; a) compulsory legislation and b) voluntary standards.

Bureau of Indian Standards (BIS)

The activities of BIS in the field of agro-processing are two fold: a) formulation of Indian standards and b) their implementation through its voluntary and third party certification system. BIS has on its record over 700 Indian Standards related to food-grains and their products, bakery and confectionery items, sugar, edible starches and their products, processed fruit and vegetable products, protein rich foods, stimulant foods like tea, coffee and coca, alcoholic beverages, spices and condiments and food products of animal origin like milk and meat, fish, poultry etc.

These standards, in general, cover raw materials permitted and their quality parameters, hygienic conditions under which the product is manufactured and packaging and labelling requirements. The standards also prescribe, where required, freedom from toxic substances and contaminants. In addition to the product specifications, which include both raw materials and final product, the Bureau of Indian Standards has brought out standards on glossary of terms for various industries and hygienic codes applicable to most of the food processing industries.

To ensure that processed foods are free from pathogenic or spoilage micro-organism, limits are gradually being introduced in various specifications. In addition, separate standard methods of test for the sensory parameters have been laid down. Containers are as important as the contents, as these may impart toxic elements to the food products and could pose potential dangers to health and safety, if they are not of the requisite quality. Indian Standards covering 6 thermoplastics namely, polypropylene isomers and ethylene acrylic acid have been published which describe the requirement of the particular thermoplastic and necessary additive along with their limits.

Informative labeling is also a very important area and the level should contain sufficient information to enable the consumers to know about positive nutritional characteristics such as protein, fat, dietary fibre etc., negative characteristics such as pesticides, residues, toxins etc. as also information regarding ingredients used, food activities, net contents etc. In this area, the Bureau has brought out a Code of Practice for labeling of Pre-packed foods covering general guidelines for labeling and guidelines on claims and nutritional labelling. However, in this area, much work still remains to be done.

Sanitary and Phytosanitary Measures

Agreement on the application of Sanitary and Phytosanitary Measures (The SPS Agreement concluded under GATT in 1994) came into effect in 1995 for developing international standards to ensure the safety of food for consumers and to prevent the spread of pests or diseases in animals and plants. These measures protect human/animal life from risks arising from additive contaminants, toxins or diseases—causing organisms in their food. The objectives of SPS can be accomplished in several ways as indicated below.

1. Requiring product to come from a disease free area
2. Inspection of products
3. Specific treatment of processing of products
4. Setting allowable maximum levels of pesticide residues or permitting the uses of only certain additives in food.

SWOT Analysis of Agro-Processing Industry

Strengths

— Round the year availability of raw materials.

— Social acceptability of agro-processing as important area and support from the central government.

— Vast network of manufacturing facilities all over the country.

— Vast domestic market.

Weaknesses

— High requirement of working capital

— Low availability of new reliable and better accuracy instruments and equipments

— Inadequate automation w.r.t. information management.

— Remuneration less attractive for talent in comparison to contemporary disciplines.

— Inadequately developed linkages between R&D labs and industry.

Opportunities

— Large crop and material base in the country due to agro-ecological variability offers vast potential for agro processing activities.

— Integration of developments in contemporary technologies such as electronics, material science, computer, bio-technology etc. offer vast scope for rapid improvement and progress.

— Opening of global markets may lead to export of our developed technologies and facilitate generation of additional income and employment opportunities.

Threats

— Competition from global players

— Loss of trained manpower to other industries and other professions due to better working conditions prevailing there may lead to further shortage of manpower.

— Rapid developments in contemporary and requirements of the industry may lead to fast obsolescence.

Plan and Strategy

The objectives of agro-processing programmes in India should be to:

- minimise product losses,
- add maximum value,
- achieve high quality standards,
- keep processing cost low,
- ensure that a fair share of added value goes to the producer

To achieve these objectives following strategy is suggested:

- National plan for improvement and extension of agro-processing technology at farm, traditional small industry and modern industry levels should be prepared. The plan should take into account the diversity in resources and needs of different regions in the Country. It should include programme details and implementation schedule for the first four or five years. The progress of plan implementation should be periodically reviewed to allow adjustments and corrective measures, and to develop programme details for the years beyond the period under review.
- Thrust areas for research and development should be identified and medium term research and development programme should be prepared and implemented to support the national plan for improvement and extension of agro-processing technology at different levels. Treatment and utilisation of effluents from agro-processing industry should be included in the R.D. programme.
- Emphasis should be put on the establishment of new agro-industrial plants in the production catchments to minimise transport cost, make use lower cost land and more abundant water supply, create employment opportunity in the rural sector and utilise process waste and by-products for feed, irrigation and manure.

— Infrastructure in the production catchments selected for agro-industrial development should be improved. Because of uncertain grid power supply to rural areas, decentralised power generation using locally available resources may become an integral part of agro-industrial development. Similarly, if the raw materials and processed products are perishable or semiperishable in nature, cold chain will have to be established.

— The national plan should provide for management of agro-industrial activities in the catchment area, both by private companies and individuals as well as cooperatives.

— Financial incentives and support should be provided on liberal scale to promote the modernisation of agro-processing industry and for establishing new such industries in production catchments.

— Arrangements to supply market information to the farmer and agro-processor should be put in place.

References

Buckmire, G., *An Overview of Small Scale Agro-Processing and Trade in the Caribbean. Constraints and Potential for Development,* "Presented to the Regional Workshop on Problems and Development Strategies for small Agro-Processing and Marketing," GUYSUCO Management Training Centre, East Coast Demerara, Guyana, 1989.

Dixon, J. A., *Technical Support Services for Agro-Industrial Development-Jamaica,* Paper presented at the Regional Consultation on Agro-Industrial Development, September 28-29, Port-of-Spain, Trinidad and Tobago, 1998.

Jones, G., *The Role of Science and Technology in Developing Countries,* Oxford University Press, London, England, 1971.

2

Seed Processing Techniques

Seed is routinely stored for more than one year, it is important to understand how seed harvesting, processing and seed storage affect the longevity and vigor of the seed. Seeds are fragile, living organisms, and the shelf life of the seed is affected at the beginning of the plant life cycle by such factors as soil nutrition. For example, if the soil is zinc deficient, the quality of the seed will be adversely affected. Though providing the best conditions for crop growth and health is the foundation of seed quality, the factors that can have the most important effect on seed viability and vigor are harvesting, extraction, cleaning, transportation, and storage. It is easy for seed to become damaged at any of these stages.

Seed Harvesting and Cleaning Methods

Seed harvesting and cleaning methods can be divided into two methods: dry processing and wet processing. Dry processing involves harvesting seed that has already matured and dried within the seedbearing portion of the plant. Examples of dry processed seed plants include beans, broccoli, corn, lettuce, okra, onions, sunflower, and turnips. Wet processing is used when the mature seed is enclosed within a fleshy fruit or berry. Examples of wet processed seed plants include cucumbers, melons, and tomatoes. Some vegetables can be either dry processed or wet processed, for example, peppers, and squash.

Rule of Harvesting

The basic rule of harvesting is to allow the seed to mature as long as possible on the plant without the seed or fruit becoming diseased, or overly ripe. Each type of plant has an optimum time for collecting the seed, but factors such as climate, weather, disease, insects, birds, or predatory mammals may require that the seed be collected at less than the optimum time. In the Mid-Atlantic and South, frequent and daily thunderstorms and high humidity may play a large role in determining how and when seed is harvested. For example, in dry climates, beans can normally be left to mature and dry in the field, but during wet humid weather, it is best to harvest early and allow the beans to continue maturing and drying under cover.

Dry Seed Processing

When seeds are ready to be processed, the entire seedpod, capsule, or seed head will become brown and dry. During the maturation process, the ripening pods and capsules change colour from green, to yellowgreen, to yellow, to light brown, to a darker brown, or dark gray. Ripening and maturation may be uneven within the pod or capsule, uneven on the plant, and uneven within the stand of plants. For that reason, the pods of many plants are harvested individually. Seeds of legumes and brassicas often develop a split along one side of the pod. This is the best time to collect the seed, before the pods start to open and scatter their seed.

Most flower seed heads are not ready to harvest until the flower head has dried completely to the base, including a short section of the supporting stem. Some plant families, such as the Asteraceae have a smaller percentage of viable seed in the head, and the seeds continue to mature after collection. For this reason it is best not to be too hasty in harvesting the seed. Examples include lettuce and sunflower. Some seed may mature in the capsule or pod, even before the pod has turned completely brown. Most seeds turn a darker colour as they mature.

Seeds may initially be white, turning green or tan, and then brown or black. Once the seed pods, capsules, and seed heads start to mature, it is important to check the crop on a daily basis. Rain or seed predators can ruin a good seed crop in a short period of time. Plants that produce umbels can usually be left in the field to harvest until the umbels are dry. Some members of this family mature their seed unevenly causing seed to scatter, while other seeds in the umbel continue to mature.

One method of dealing with crops that mature their seed unevenly is to pull the plants and hang them upside down to dry under cover. This allows the seed to continue to mature on the plant while the plant dries. This procedure is often used for lettuce. Confidence in knowing when to harvest comes both with experience and familiarity with different species and crops. After harvest, seeds are threshed to remove the seed from the surrounding plant material. A period of air-drying is important before seeds are threshed. Plant material should be spread out in thin layers until all plant material is dry; otherwise, mold, decay, and heat from decay will cause damage to the seeds.

As the plant material dries, seed pods may split open or shed seed. Harvested material should be stored in a well ventilated room with low humidity. During this time you should be aware of insects, especially weevils that feed on the seeds. Plant material that is ready to be threshed should be brittle. Threshing is best done outside on a dry day. The threshing process involves application of mechanical force using a controlled pressure and a shearing motion., and is accomplished by hand or by machine. There are many different methods for threshing seed.

Plants that have pods, such as beans and okra, can be threshed by placing the pods in a large feed sack, which is tied shut securely, and then placed on the ground where it is flailed, stepped on, jogged on, or danced on with a twisting motion. The sack is turned often to redistribute the plant material for further threshing. When using this method, it is best to use running shoes or other soft-soled shoes because seeds can develop hairline cracks and splits from too much pressure.

To separate seed from flower heads, plant material is spread on a concrete slab and then gently walked on with a twisting motion to break open the flower heads. Care must be taken to not apply so much pressure that the seed is abraded or broken. To thresh seeds of brassicas, place plant material in a large wheelbarrow or on a large tarp, and while wearing gloves twist and wring the plant material through your hands until the seed breaks free from the pods. Another method for extracting seed is to place seed heads on a piece of plywood.

The seed is extracted by placing a wooden cement float above the material and then pressing and twisting until the seed breaks free. If the seed heads are small you can run a rolling pin over them. For threshing small lots of seed, a threshing box may be used. This consists of a wooden box, with sides slanting outward, open at the top and on one end, and the long sides and bottom covered with corrugated-rubber floor matting. Seed is placed in the box and a rubbing board, is moved back and forth across the seed heads to separate the seed from the heads.

Wet Seed Processing

Wet seed processing is used with seed crops that have seeds in fleshy fruits or berries. There are three steps to the process:

(1) extraction of the seed from the fruit,

(2) washing the seeds, and

(3) drying.

Extracting Seed

The type of extraction process depends on the species. Soft fruits such as tomatoes are cut up, mashed, and then fermented. Cucumbers and melons are cut in half, the seed scraped out along with the fruit pulp surrounding the seed, and then fermented. In watermelons, the entire fleshy fruit is fermented along with the extracted seed. These types of fruits have a gel surrounding the seed that contains germination inhibitors. The presence of the gel also makes handling and drying of the seed difficult. Fermentation is a natural process that occurs to a small extent as fruits decompose. When fermentation is done in a controlled manner, the microorganisms, principally yeast, break down the gel thus releasing the seed while killing bacteria and fungi that cause most seed-borne diseases.

The temperature and length of fermentation are important. If the mash is not fermented long enough, seedborne diseases will not be eliminated, but if fermented too long, the seeds may sprout prematurely. The length of the fermentation is dependent on temperature and typically last three days at a temperature of 70 to 75°F (21°C to 24°C). Length of fermentation may also depend on the variety itself. For example, varieties with high sugar content may take longer to ferment, up to four days. With few exceptions, fermentation periods longer than three days risk damaging the seed.

There are different fermentation techniques for different crops, for example, pepper seeds are extracted from the fruits by mashing, but the fermentation process may last only 24 to 48 hours. Though eggplant isn't a watery fruit, it can be mashed and fermented for about 48 hours. A Corona® grain mill is a very handy tool to use for extracting seed of eggplants and hot peppers. The grinding plates on the mill are set wide enough apart to break up the fruit without grinding the seed.

Small peppers can be put directly into the mill, whereas larger fruits such as eggplant are cut into small cubes before processing. Before processing another variety, the mill has to be disassembled and cleaned thoroughly to remove bits of small seed caught in the plates and other parts of the mill. Though some people use a food processor and blender set on low speed to extract the seed of hot peppers and eggplant, it is easy to damage the seed.

Some varieties make a very thick mash that is hard to stir, and others make a watery mash that stirs easily. The mash should be stirred three times a day, once in the morning, once in midday, and once in the evening. If the mash is too thick to stir easily, the nutrients are not going to circulate easily.

Stirring is also important for better control of seedborne diseases. In addition, when the mash is not stirred, a foul-smelling, white mold forms on top of the mash. This mold can discolour (darken) and damage seeds at the top of the mash. These seeds will later have to be hand picked out of the dried seed. A properly fermenting mash should not have a foul smell, and there should be little or no white mold on the top. A small amount of white mold is not harmful and can be stirred back into the mash, but a heavy overgrowth should be removed.

Washing Seed

After fermentation is complete, the seeds are washed to remove pulp, pieces of fruit and debris, and low quality seed. Before washing the seed, it is useful to first scoop out pieces of pulp floating on top of the mash. This is done by straining the mash with your fingers, pulling out the larger chunks. Whether or not there is floating pulp depends on the variety or how thoroughly the fruit was processed. Add a volume of water at least equal twice the volume of mash. It is important to dilute the mash sufficiently because the more dissolved solids there are in the mash, the higher the specific gravity.

If the specific gravity is high (lots of soluble solids) it will be more difficult to wash the seeds properly. As a general rule, good seeds are heavy and sink to the bottom, whereas poor quality seeds are light and tend to float off with the wash. The washing process is repeated until the wash water becomes clear. Although most good seeds sink to the bottom, some vegetables have very light seed and require extra care during the washing process. For example, it is common for a significant amount of good pepper seed to float rather than sink during the washing.

This can be avoided by adding the wash water slowly, so as not to create tiny air bubbles that adhere to the seeds, making them buoyant. Even with this precaution, there are a few varieties of watermelon for example, where the good seeds tend to float rather than sink, thereby requiring special care in washing. In this case, the best way to wash the seed is to pour the wash through a ¼" hardware cloth screen, and then use a hose to force pulp through the screen.

Drying Seed

Seeds should be dried fairly quickly after washing. Slow drying may result in mold growth or premature sprouting of the seed. In the Mid-Atlantic and South, seeds should not be dried in the sun, nor should they be dried anywhere where the temperature exceeds 95°F (35°C). Dark coloured seeds are especially vulnerable to damage when sun dried. Instead, seeds should be dried in a climate-controlled environment using fan ventilation.

A combination of ceiling fans and air conditioning dries seed safely and very quickly. Seed should be spread out in thin layers and then stirred several times a day until dry. Once the seeds feel dry, they should cure for another two to three weeks. Curing is the final stage in the drying process. As the seed moisture content declines it comes into equilibrium with the relative humidity. After the seeds are cured they can be placed in a container.

When drying seeds, choose plywood, window screen, or any hard, non-stick surface. Avoid using paper towels, newspaper, cardboard, or cloth because seeds will tend to stick to the surface making them difficult to remove. Beginning growers often make the mistake of drying squash seed on newspaper, which adheres permanently to the seed coat. No one wants to read the daily news on the surface of squash seed.

Seed Processing Equipment

Threshing Equipment

For small-scale seed production, it is often not cost effective to purchase threshing equipment because of the small volume of seed produced. A number of small alternative seed companies, specialising in heirlooms and specialty seed, thresh most, if not all of their seed by hand. Some seed crops, especially herb and flowers seeds are produced in such small quantity (and have such small seeds) that it is impractical to use a machine for threshing. Also flowers and other seed crops, if left to dry on canvas or a tarp, naturally dehisce (shed) much of their seed.

The remaining amount of seed can be released by flailing or rolling, or other non-mechanical means. It is difficult to locate low-cost, low-tech shredders for small-scale seed threshing. The options are to locate old, used equipment or to construct your own. Vegetable seeds, such as beans can be threshed with a modified leaf shredder/chipper.

Typical threshing rates for this machine are 22 to 30 pounds/hour for brassicas, and approximately 100 pounds/hour for beans. The research and development group at the Organic Gardening Experimental Farm of Rodale Press developed a small-scale thresher designed to thresh a variety of crops ranging from amaranth to soybeans. Though the machine was not fully refined it did an acceptable job. Plans for this machine were published in Small Scale Grain Raising.

Seed Cleaning Equipment

It is not necessary to have expensive seed cleaning equipment to clean seed for small-scale production. With the exception of lettuce seed and a few others, the majority of crop seeds can be cleaned with homemade seed screens. Winnowing will still be necessary to remove smaller chaff. Many seeds can be screened with several different mesh sizes of hardware cloth. Hardware cloth is readily available in the following mesh sizes: ½", 3/8", ¼", and 1/8". The ½" and ¼" mesh sizes are available at most hardware stores. The 3/8" and 1/8" mesh sizes are the most useful sizes, but are the most difficult to find.

Aside from hardware cloth, a lot of other materials are useful for making seed screens. Aluminum window screen can be used for small seeds. You'll also find various meshes of screen available in the housewares section of department stores. Cabinets from electronic devices often have round or oblong holes which are useful for cleaning some types of seed. Special meshes can be ordered from mail-order hardware specialty catalogs. It is useful to collect a large variety of mesh sizes and shapes to handle a wide variety of seed types. Once you have located suitable screen material it should be mounted on a frame. Professional seed screens are mounted on 12" square wooden frames.

Homemade screens should also be mounted on 12" square frames so they can be used together with professional screens. The frames can be made of wooden lath which measures ¾" x 1-1/2". Ideally, the screens should be constructed so that they nest together. The nesting feature is desirable for using three screens simultaneously: rough chaff is retained on the top screen, seeds in the middle screen, and small chaff and small seeds on the lower screen. Once the frame has been made, the screen is nailed with small nails or brads at approximately 1" intervals all along the edge of the frame.

Equipment for Winnowing

The classic method of winnowing involves placing seeds in a wide basket and tossing the seeds and chaff into the air. The chaff is carried away by the wind. Illustrations of Native Americans using this historic method evoke a bucolic mood, but in reality, winnowing by this method is extremely difficult, and the results are not very satisfactory. The most vexing part of the process is that wind speed is always changing in velocity and direction. It can work for certain kinds of seed,. but it actually works better to use two large bowls, pouring seed from one bowl into another bowl below, while blowing on the plant material as it falls. This method works satisfactorily if the seed is heavy and the chaff is very fine, and susceptible to being carried away by a gentle current of air. In any case, hand winnowing should be done, not on windy days, but when the air is calm.

A good range of equipment for winnowing includes the following:

(1) an assortment of stainless steel bowls ranging in size from 6 to 16" in diameter, preferably bowls with varying shallow and deep sides;

(2) an electric hair dryer with two speeds (heating element removed, or unheated air setting);

(3) portable vacuum cleaner with option of connecting hose to the air discharge opening:

(4) household box fan with three speeds;

(5) squirrel cage blower;

(6) rheostat for controlling fan motor speed;

(7) a tarp for catching seed or chaff; and

(8) a dust mask for keeping chaff out of your lungs.

Most light seed can be winnowed quickly and efficiently with mixing bowls and a hair dryer. It takes a little practice determining the correct amount of seed to put in the bowl, setting the proper fan speed, and the distance of the dryer from the bowl. It helps to jiggle the bowl up and down or to swirl the seed in the bowl. There must be little or no wind when you work with light seed. Until you have a little practice it is best to put a tarp on the ground, in case you need to sweep up your mistake and start over.

Heavy seed is best winnowed by pouring the seed from one container to another in front of box fan, or alternatively directing the air discharge hose from a vacuum cleaner into a large (16" diameter) mixing bowl. When dealing with large volumes or certain types of seed it is helpful to use mechanical equipment for winnowing. The University of California at Davis has plans for making your own winnowing equipment. This information is available from their Long Term Research on Agricultural Systems project dealing with Appropriate Technology for Small and Subsistence Farms.

Seed Drying Process

Principles

Drying is a normal part of the seed maturation process. Some seeds must dry down to minimum moisture content before they can germinate. Low seed moisture content is a pre-requisite for long-term storage, and is the most important factor affecting longevity. Seeds lose viability and vigor during processing and storage mainly because of high seed moisture content. High seed moisture causes a number of problems:

— Moisture increases the respiration rate of seeds, which in turn raises seed temperature. For example, in large-scale commercial seed storage, respiring seeds may generate enough heat to kill the seeds quickly, or to even start a fire if not dried sufficiently. Small-scale growers are not likely to have such an extreme condition, but seed longevity will, nevertheless be affected.

— Mold growth will be encouraged by moisture, damaging the seeds either slowly or quickly, depending on the moisture content of the seeds. Some molds that don't grow well at room temperature may grow well at low temperatures causing damage to refrigerated seeds. In such a case there may be no visual sign of damage.

— Unless seed moisture is at least eight percent or below, insects such as weevils can breed causing rapid destruction of seeds in a short period of time.

Drying for Longterm Storage

Silica gel is the most effective desiccant (moisture absorbing material) for drying seeds. Powdered milk has been recommended as a desiccant in older seed-saving literature, but is less than ten percent as effective as silica gel. Silica gel is a highly porous form of silica that absorbs moisture. It is available as a powder or as beads in different sizes.

The best size bead for drying seeds is approximately 1/16" to 1/8" in diameter "Colour-indicating" silica gel is a form of silica gel that has been treated with a small amount of cobalt chloride which acts as a moisture indicator. When the indicator gel is completely dry it is a deep blue colour. As the gel absorbs moisture from the air, it gradually changes in colour from deep blue to light pink. Though silica gel is clearly allowable under the Organic Rules, colour-indicating silica gel is not clearly covered. Check with your certifier before using colour-indicating silica gel.

Silica gel can be repeatedly reactivated (re-dried) after it has absorbed moisture. The procedure involves heating the gel and driving off the moisture, and as it dries the colour gradually changes to deep blue. Drying must be done in a controlled fashion, otherwise the beads will turn black, and the moisture-absorbing capacity of the beads will be destroyed. There are two methods for reactivating silica gel:

Oven-drying Method

This method gives the best results, but it takes longer and uses more energy. Set the oven for a temperature of at least 200°F, but no higher than 275°F. Remember, that some ovens may run hotter than the set temperature. Place the silica gel in a thick-walled Pyrex dish, no deeper than one inch, and continue heating until the beads turn deep blue. When drying large quantities (a pound or more), the gel should be stirred occasionally. The oven drying method takes 1-1/2 hours per quart of gel (with the oven temperature set at 275°F). One quart of silica gel weighs approximately 30 ounces (1.9 pounds).

Microwave Method

The microwave method works much faster but must be monitored more closely to avoid overheating the gel. Use only a thick-walled Pyrex container for heating the gel. Set the microwave on medium or medium high and dry for approximately three to five minutes. The colour change in the gel can be monitored through the over door, but the gel should be inspected and stirred at the end of each heating cycle. If the gel has not dried, heat again for another three to five minutes.

Approximate drying time is eight to twelve minutes per pound of gel, though actual heating time will depend according to the type of microwave. There are some fine points and cautionary notes about using silica gel. When using a Pyrex dish, the glass should be thick. Do not use plastic microwave containers that will melt on contact with the gel. Silica gel gets very hot, and the glass container may shatter if it is too thin, of the wrong type, or if unevenly heated. You may notice a slight odour during heating.

This is due to either overheating the silica gel, or to the evaporation of organic seed volatiles absorbed by the silica gel during the drying process. Silica gel itself is chemically inert, non-toxic, non-corrosive, and odourless, but breathing the dust can be hazardous under prolonged and repeated exposure. For that reason, do not use the finely ground or powdered form.

Drying Seed with Silica Gel

— To dry seed, determine the weight of the seed to be dried, including the packets or envelopes that contain the seed. Measure out an equal weight of silica gel and place the seeds and silica gel in an airtight container for seven days. When drying seed it is important to keep the container size small in relation to the volume of seeds being dried.

— At the end of seven days, remove the packets of seed from the drying container and transfer into another airtight container, such as a Mason jar, Seed Saver Vial™ or barrier pouch. Because seeds can re-absorb moisture from the air quickly, they should be transferred quickly.. When used as directed, silica gel dries seed from 12% typical moisture content to a desired moisture content of approximately 5% for small seeds and 7% for large seeds. As a "rule of thumb", seeds will not be damaged provided the drying time doesn't exceed seven days. This drying time applies to humid climates, such as the Mid-Atlantic. Longer drying may drop the moisture content below 3% for small seeds and below 5% for large seeds, levels which may damage seeds or force them into dormancy. Legumes, such as beans, are especially are injured by over drying.

Factors Affecting on Seed Longevity

Effect of Temperature

The general effect of temperature on longevity is that longevity increases as temperature decreases. This is true of "orthodox" seeds: that is, most seeds that follow some general "rules of thumb" regarding longevity during the storage life of the seeds. The relationship between temperature and seed longevity is that for each 10°F (5.6°C) decrease in temperature, longevity doubles. This rule applies to seeds stored between temperatures of 32°F (0°C) and 122°F (50°C).

This rule assumes that the moisture content is a constant. This is a general guideline; in reality the longevity of some vegetable species declines more rapidly than suggested by the rule, while the longevity of others declines more slowly in relation to storage temperature. The longevity of seeds is generally not affected by subfreezing temperatures provided the moisture content is less than 14% (because ice crystals do not form). This is the ideal way to store seed, especially small seed that doesn't require much freezer space. One caveat: seed cycled in and out of the freezer too many times without redrying may cause degradation of germination.

Effect of Seed Moisture and Humidity

Seed moisture has a greater effect than temperature on seed longevity. Most seeds also follow some "rules of thumb" regarding moisture and longevity. The general relationship is that for each one percent increase in seed moisture, longevity decreases by half. This rule applies to seed with moisture content between 5 and 13%. Above 13% moisture content, seed storage fungi and increased heating due to respiration cause longevity to decline at a faster rate.

Once seed moisture reaches 18 to 20%, increased respiration, and the activity of microorganisms cause rapid deterioration of the seed. At 30% moisture content, most non-dormant seeds germinate. At the low end of the moisture range, seed stored at 4 to 5% moisture content is unaffected by seed storage fungi, but such seeds have a shorter longevity than seed stored at a slightly higher moisture content.

Effect of Relative Humidity and Seed Moisture Content

When storing commercially grown seed, it is impractical and too costly to use desiccant to dry the seed for storage, unless the seed is small and expensive. Commercial seed is usually packaged for short or longterm storage under conditions of ambient humidity (unless special equipment is used). Because relative humidity has a significant effect on seed moisture content, it is important to understand the relationship between humidity and seed moisture. Regardless of the type of storage conditions, the moisture content of seed eventually comes into equilibrium with the moisture in the surrounding air. The relationship between atmospheric relative humidity and seed moisture content is shown at the right in Figure 1.

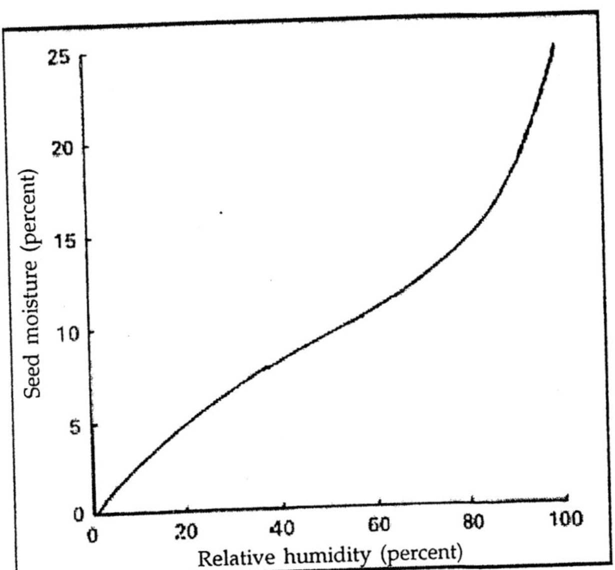

Figure 1. Relationship of vegetable seed moisture content to relative humidity.

The curve was derived from measurements of the average seed moisture content of ten vegetable species stored at different relative humidity. Because the curve represents an average of ten different vegetable crop species, the response of individual species may vary. For example, seeds of grains (which contain relatively high percentages of carbohydrate) will have a moisture content of 13 to 15% at 75% relative humidity, whereas seeds rich in oils (such as peanuts) can have a moisture content of 9 to 11% at the same humidity.

Note that once the relative humidity reaches 70%, the moisture content of the seed has reached approximately 13%, the point at which increased respiration and seed storage fungi become a significant problem. Above 70% relative humidity the moisture content rises dramatically. In central Virginia, the relative humidity in a residence or office building during the summer usually averages about 65%, plus or minus 5%. If air conditioning is used, the relative humidity typically ranges between 50 to 60%, depending on how often it runs.

During the winter heating season, the relative humidity averages about 40%, plus or minus 5%. If the seed is stored in an open barn, or outbuilding, the relative humidity surrounding the seeds will be higher than that of a climate-controlled dwelling. or office building. By using the values above, and referring to the chart above, it is possible to get an estimate of the moisture content of stored seed. It is clear from looking at the chart, that it is important to store seeds in a climate-controlled environment, especially during the summer.

Seed Moisture Content

Moisture content of seed is defined by the International Seed Testing Association (ISTA) according to the following formula:

$$\textit{Seed moisture content}\,(\%) = \frac{\textit{Fresh seed weight} - \textit{dry seed weight}}{\textit{dry seed weight}} \times 100$$

In order to determine the percent moisture content of fresh seed, a sample of fresh seed is weighed, and then an equal weight of fresh seed is dried slowly to remove the moisture, and then re-weighed. The equipment used to do this is a "seed moisture balance", which has an infrared heating lamp mounted above the balance pan. The fresh seed is weighed in the balance pan, and then the lamp is turned on. The weight of the seed gradually decreases during heating until the weight decreases no further.

Once the weight comes to equilibrium under the heating lamp, the dry weight is recorded. The data is then used in the formula above. A seed moisture balance is an expensive piece of equipment, but reliable information can be also obtained by heating the seed slowly in a toaster oven at low heat until dry, and then reweighed. Using this method it may be necessary to weigh the seed several times during drying to determine if the weight has come to equilibrium.

Relationship between Temperature and Moisture

The effects of temperature, moisture, and relative humidity were discussed above as separate factors which affect the longevity of stored seed. In reality, the effects of temperature and relative humidity are highly interdependent in their effect on stored seed. There is a simple method for calculating the combined effects of relative humidity and temperature on seed longevity, which is as follows: the sum of the storage temperature (in degrees F), plus the relative humidity (in percent) should not exceed 100.

Since seed moisture is the most important concern, the rule stipulates that no more than half the sum should be contributed by the temperature. The majority of crop seeds lose viability quickly when the humidity approaches 80% at temperatures of 77°F (25°C) to 86 F (30°C), but when stored at a relative humidity of 50% or less, and a temperature below 41°F (5°C), seeds will remain viable for at least ten years.

If seeds are taken from a cold or frozen storage and transferred to room temperate, care must be taken to prevent condensation on the seeds. If the seeds are in a sealed container, allow them to sit until they reach room temperature before opening the container. If they are stored in paper, place the seeds into a plastic bag with the excess air sucked out, seal the bag,, and wait for the temperature to stabilize before unsealing.

Effects of Light

The effects of light on stored seed have been studied: Some studies showed a benefit and some showed a detriment: the results are inconclusive and controversial. Seeds stored in glass containers should be stored out of direct sunlight because of the localised "greenhouse heating effect" on seeds. This might seem an unnecessary and overly obvious cautionary note, but in my experience, it has happened accidentally more than once. For example, this can happen if jars of stock seed are taken outside for the purpose of removal of some seeds for planting, or for transporting to another location.

Though the jar may be temporarily stored in the shade, the angle of sunlight may change quickly and the jar will be in direct sun causing very rapid heating within the jar. Though some commercially produced seed is dried in direct sunlight (in dry climates), drying seeds in the sun is a questionable practice in the Mid-Atlantic and South if the air temperature is above 90°F (32°C).

The air temperature at the seed surface is higher because of the conversion of light energy into heat at the seed surface, and the heat is "moist heat" (though this wouldn't be in an issue in dry climates where evaporative cooling occurs at the surface). Another concern, also not well documented, is that the ultra-violet light from the sun may have a deleterious effect on seed longevity while the seeds are drying. Harrington's suggestion was based on the known effects of ultra-violet radiation on biological systems (rather than specific data). Whether the ultra-violet exposure is long enough to cause concern, is unknown.

Effect of Respiration and Heating

The largest factor affecting respiration and heating is moisture content, and therefore, at a minimum, seeds need to be kept dry. Respiration in seed storage has three effects:

— Depletion of food reserves. Over the life of stored seed, depletion of food reserves is inconsequential.

— Release and accumulation of gasses that may affect viability of seeds in storage. The accumulation of respiratory end products such as carbon dioxide is an advantage of storing seeds in sealed containers where carbon dioxide replaces oxygen in the air thereby slowing down respiration and increasing longevity.

— Release of energy, mostly in the form of heat. When seeds are stored under favourable conditions, respiration is of little consequence. When moisture is high, respiration increases, which in turn increases the production of heat thereby decreasing longevity.

Effects of Fungi, Bacteria, and Pests

The process of seed harvest and cleaning removes most debris and insects, but certain fungi, bacteria, and insects make their way into stored seed. Fortunately, the same conditions that are favourable to seed preservation inhibit fungi and bacteria and kill insects.

— *Bacteria*: Bacteria do not have a significant role in seed deterioration because free water is required for bacterial growth, and if the moisture content of the seed is high enough to support bacteria, the seed is more likely to succumb to deterioration due to other causes such as fungi, respiration, heating or premature sprouting.

— *Fungi*: Most seed storage fungi are inhibited when the relative humidity is kept below 65%. At this relative humidity the moisture content of starchy seeds is about 13%, and oily seeds about 7%. The major effects of fungi are to:

— decrease viability;

— produce toxins that affect seed viability and germination;

— increase heat production - important in large seed lots; and,

— cause discoloration, mustiness, and caking.

— *Insects*: In hot, humid climates such as the Mid-Atlantic and South, mites, weevils, flour beetles, and borers can be a serious problem in stored seed, but if the seed is dried to 8% moisture content and the temperature reduced to (64 to 68°F (18 to 20°C), insects should not be a problem. At a moisture content of 15% and a temperature of 86 to 95°F (30 to 35°C), they can become very destructive. Mites will not survive when the relative humidity is below 60%.

Different Varieties and Harvest Conditions

Different varieties of a particular crop species may have different longevity when stored under the same conditions. This may be explained by differences in enzyme activity, or differences in the chemical constituents of unique varieties. For example, sweet corns are rich in simple sugars, whereas dent corns are high in complex carbohydrates. This explains why dent corn seed keeps longer than seed of sweet corn. Harvest conditions may also affect longevity of seeds in storage. For example, under poor conditions of harvest, seed may have more fungi on the seed coat. Differences in harvest conditions are insignificant if storage conditions are good, but under adverse conditions, harvest differences can be very significant.

Seed Moisture Levels

Seed can be stored for long-term sealed storage provided that the seed moisture content is less than 8%, which means that the relative humidity must be kept below 35%. For the Mid-Atlantic and South, seed must be dried with a desiccant in order to achieve the desired moisture level for long-term storage. When drying with silica gel according to the procedure described above, large seeds such as peas, beans, and corn will be dried to about 6 to 8% moisture content, and smaller seeds will dry down to about 3 to 5% moisture content, depending on the size of the seed and whether its primary food reserve is starch or oil. Seeds with a starchy food reserve retain a higher seed moisture content than oily seeds.

Packaging Materials and Containers

Shortterm Storage Materials

There are a wide variety of materials that can be used to store seed for short-term storage. Most of these are non-rigid materials such as cotton, burlap, paper, and composite materials such as multi-wall paper and plastic film, or polyethylene bags. Materials used for short-term storage are generally porous. They adequately contain and protect the seeds from mixing, but do not provide protection from moisture or loss of seed viability. Such materials are usually used for mechanically separating seed lots, and for transporting and shipping seed until the seed can be placed in environmentally controlled conditions for longer-term storage.

Each type of packaging material has its own advantages and disadvantages. Burlap bags have the greatest strength, can be re-used many times, and can be stacked high without slipping of the stack. The strength of cotton bags is dependent on the thickness of the weave and thread, and the quality of the seams. Though not as strong or as tear-resistant as burlap, cotton bags can often be re-used, depending on the quality of the fabric. Bags made of woven plastic material also are fairly strong but tend to slip when piled high and are harder to close securely after the bag has been opened.

Multi-wall bags are made of several layers of paper in a variety of types of construction. They have poor bursting strength when piled high or accidentally dropped, and if used repeatedly, they tend to become brittle along the folds and wear points. Cardboard boxes and cans, though expensive, are re-usable, good for stacking, and provide some protection against mechanical injury to seed and to infestation by seed storage insects. Flexible packaging that has a weave, whether it is burlap, cotton, or plastic offers little protection against seed storage pests such as grain moths.

Consequently, seed stored in such woven bags may have to be inspected if an outbreak is detected in a particular seed bag. Seed storage insects are very good at locating small openings in bags or containers that are not well sealed. For small lots of seed, paper bags such as lunch bags are inexpensive and adequate for storing seed, but the seams are not always reliable, and when used, the bags should be double or triple bagged to ensure integrity to prevent bursting. Other materials such cellophane, acetate, and 2- to 4-mil polyethylene zip lock bags may be used. If using polyethylene, it is best to use the 4-mil thickness, especially for heavy seeds. Plastics and thin films are not reliable moisture barriers, though such materials offer better moisture protection than paper.

Longterm Storage Materials

Metal and glass containers, properly sealed to prevent the exchange of moisture and gas, are the most commonly used containers. They are the only reliable means of protecting seeds against humidity, insects, rodents, floods, and mechanical damage. Plastic should not be used for long-term storage. For storing large quantities of seed, metal five-gallon cans fitted with a rubber gasketed lid and pressure ring are ideal for storing large seeds such as peas, beans, and corn.

One-gallon jars are also excellent, provided that the lid has a gasketed seal. Some one-gallon jars have plastic lids that are flexible enough to form a tight seal when the lid is screwed tightly, but metal lids need a rubber gasket. Gaskets can be cut from sheets of neoprene rubber (available at hardware stores) or used automobile inner tubes. Though glass is breakable, it has the advantage of being transparent so that the contents can be easily inspected for insect damage (especially useful for detecting bean weevils, grain moths, and other seed pests).

Glass canning jars are available in sizes ranging from ¼ pint to ½ gallon. Onegallon glass jars are often available free for the asking at restaurants, fast-food restaurants, and snack bars that purchase pickles, relish, mayonnaise, and a number of other food ingredients in bulk. For storing small quantities of seed, ¼ and ½ pint jars are ideal, provided the rubber seal is in good condition. Baby food jars are often recommended, but again the seal should be in good condition. The seal can be ruined from seeds getting trapped between the rim of the jar and the lid; therefore jars should not be packed to the rim. Other materials that can be used are Seed Saver Vials™which are made of a high-density polypropylene with a patented inner and outer valve that ensures an airtight seal. Another material, often used for seed banks, is a heat-sealable barrier pouch. Barrier pouches are a triple-laminate material made from paper, plastic, and foil.

When deciding on the type of storage container, it is often helpful to settle on a minimum number of standard sizes. If using canning jars, it is best to choose a particular brand name and style so that same-size and same-style containers have the same weight. This is very helpful in inventory control of seeds being held for long-term storage. Standardisation of container style and weight makes it easy to obtain the weight of stored seed without actually removing the seed from the container and exposing to humid air. Once the weight of standard containers is known, it is easy to subtract that weight from the container of seed to obtain the weight of the seed.

Seed Longevity

Seed Viability and Vigor

Seed germination, generally measured by percentage, measures the number of seeds in a lot that can be expected to germinate and grow healthy plants. Seed vigor is defined by normal seedling morphology and the rate at which seeds germinate and grow in the early stages. Strong seed vigor has many advantages for the organic grower, as vigorous seeds are less likely to be overtaken by diseases, weeds, and insects than weak ones.

During the time that seed is held in storage, there is a gradual decline in germination and vigor. When a seed company holds seed in storage at ambient temperature, the seed is typically tested every six months to a year, depending on the conditions of storage. Seed that is stored at 40°F (14°C), or at subfreezing temperatures is tested on a longer cycle depending on how the seed was dried and the temperature at which it was stored. Every time a germination test is done, the germination percent and test date is recorded on the stock container. What that label doesn't show is the degree of seedling vigor. There should be a germination test number on the label so that the test log can be consulted if there are concerns about vigor.

Loss of seedling vigor is often apparent in the germination test. When the seeds take longer to germinate than usual, the seedlings are smaller, and sometimes malformed. Information about the vigor of a variety can be partially inferred from the germination data if you are familiar with the general relationship between viability and vigor. The relationship between vigor and viability is illustrated by a sigmoid survival curve as shown in Figure 2.

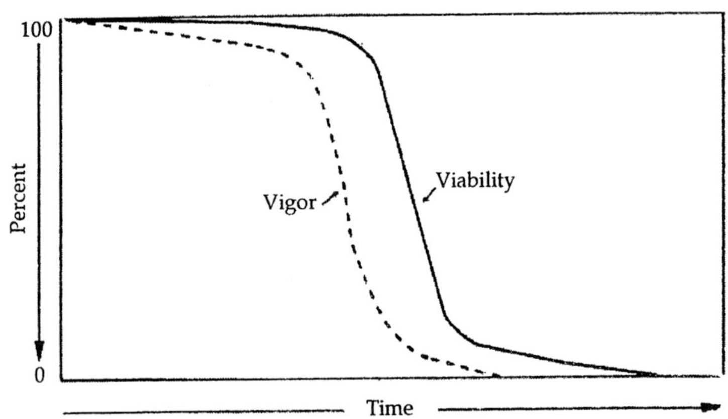

Figure 2. Relationship between the decline in vigor and viability over time.

Note that the relationship between vigor and viability is similar except that vigor declines before viability. There are three distinct phases to this relationship:

(1) the first when germination is approximately 80% or above, when seed is both vigorous and viable;

(2) the second stage when deterioration progresses rapidly; and

(3) the third stage when deterioration slows at approximately 20% or below, and all seeds slowly die.

The curve shown in this graph is a generalised curve, and that there are significant differences among varieties and types of seeds in regard to viability and vigor. Large differences in vigor and viability are more likely to show up under substandard storage conditions, especially when the relative humidity permits the growth of storage molds.

Seed Dormancy

Every type of seed goes through a period of dormancy, which is a mechanism for delaying germination until the seed is dispersed and exposed to favourable growing conditions. It is not evolutionarily advantageous for all seeds from the mother plant to germinate at the same time. Environmental conditions, such as drought, disease, and predation may wipe out a generation of seed.

Dormancy provides a mechanism for seeds to be able to adapt to their environment by germinating at different times and under different conditions. It is not a mere resting state of a seed, but rather an active physiological state. Dormancy helps seeds to germinate at the proper time of year, temperature, pH, lighting conditions, nutrient levels, or other environmental conditions that affect survival. It allows weed seeds to remain ungerminated in the soil for many years, until the conditions are conducive to growth.

Most vegetable crops have a long history of domestication, a selection process which has eliminated or reduced dormancy in many of our major food crops. Nevertheless, some of our common crops still exhibit some dormancy, and this poses a problem for seedsmen. For example, some varieties of brassicas, tomatoes, peppers, especially varieties with wild ancestry, may not germinate well, or at all, shortly after harvest. A period of dry storage may be required before the seeds germinate. The process of artificially breaking dormancy is called conditioning. For 95% of seeds, the basis of dormancy is biochemical or physiological in nature, whereas for the remainder of dormant seeds, the cause for dormancy is an impermeable seed coat.

The mechanism for breaking dormancy varies from species to species (and even variety to variety) and most often involves drying (curing), exposure to light, leaching of chemical inhibitors, exposure to high or low temperature, or alternating temperatures. The germination rate of many agricultural seeds may be increased if exposed to alternations in daily temperature. For example, some herb seeds tested in the controlled conditions of a germination chamber do not germinate as well as seeds germinated outdoors in the same media. In some species, dormancy is broken by exposure to nitrate ions in the soil, and in others, exposure of seeds to gibberellins (plant hormone) produced by soil fungi.

Seeds that are produced by fleshy fruits have dormancy broken by fermentation and leaching of the chemical inhibitors from the gelatinous coat surrounding the seed. Preharvest growing conditions can also affect dormancy. For example, lettuce grown under different conditions of water and nutrients exhibit differences in the amount of dormancy. Squash and pumpkins have the highest germination percentage when seed is harvested from fruit that is twenty days past peak ripeness. In many instances, seed dormancy is released by dry storage.

The seeds of crop species that exhibit dormancy after harvest can be released from dormancy within two to six months after harvest. This is important for some pepper varieties. If the seeds are frozen, the seeds remain dormant until they are subjected to above freezing temperatures for several months. It is important for seed growers and seed purveyors to understand the dormancy characteristics of crops, especially in regard to germination testing. A seed grower testing the germination of a lot of freshly harvested seed may be surprised to discover that the tested seed doesn't germinate. Rather than assuming that seed is dead and needs to be discarded, it is important to test the possibility that the seed is dormant, and that the seed may need to be retested after a period of conditioning.

STATE SEED LAWS

Seed laws exist for the basic purpose of ensuring that seed is labelled truthfully. They serve to protect gardeners and farmers, and because seeds move from dealer to dealer they protect seedsmen as well. Effectively, anyone who handles seed in commerce is afforded protection of the law. When a particular variety is in short supply, seed dealers may legally sell low-quality seed, but the seed must be labelled as "below standard", and additional seed may need to be added to the container to compensate for the low germination.

In most states, seed packets are not required to show the actual percentage germination unless the package weighs a pound or more. Other labelling requirements exist for some crops, especially grains, indicating the percentage of weed seed. Some crops have hard seeds - seeds that are impermeable to water during the usual germination period. When these crops are labelled, the percentage of hard seeds is included in the germination count. Another protection afforded by labelling is the specification that seeds that have been treated must be labelled as such.

In addition, a dye must be added to the seed so that it can visually be identified as treated seed. Seed, that has been grown organically, must meet organic certification standards before the seed can be labelled as organic seed. When growing seed for sale, you should contact the Department of Agriculture in your state to obtain a copy of the seed laws for your state. Certain crops such as cotton may require a permit to grow the crop. In Virginia for example, the state monitors boll weevils by placing traps in the field. Other states control the sale of seeds, which may be considered noxious weeds, and either forbids their sale or their importation into the state.

State seed laws contain fairly similar requirements, so if you are meeting the requirements of the Federal Seed Act, very likely you are in compliance with state seed laws. Some states have different germination minimums for crops, but these differences pertain more to herbs and flowers than they do to vegetable seed. Federal germination standards are specified in the Federal Seed Act. Some crops, such as peppers have a minimum germination standard of only 55%.

Though this seems low, when dealing with varieties of peppers that have wild genes in their ancestry, germination in these varieties is lower and dormancy tends to be an issue. In fact, the 1966 Virginia Seed Law gives a standard of 75% germination for a list of older varieties and a standard of 70% for newer varieties. Seed companies generally require their growers to produce seed with higher germination rate than that of federal standards because they need seeds with a long shelf life.

Ideal Seed Treatments

The ideal seed treatment should be:

(1) very effective against seedborne pathogens,
(2) relatively nontoxic to animals and plants, even if misused,
(3) effective for a long time during seed storage,
(4) easy to use,
(5) acceptable according to the OMRI list of practices, and
(6) economical.

The hot water treatment method meets many of these criteria. Though not as easy to use as chemical treatments, it can be more effective and is non-toxic.

The purpose of the hot water treatment method is to submerse seed in water hot enough to kill the pathogen without damaging the seed embryo. When properly done the procedure is very effective, but there are some precautions to follow:

(1) use only new, vigorous, high germination seed,
(2) control the temperature of the process very carefully, since this is fundamental to success,
(3) test the germination of the seed lot before and after treating the seeds, and

(4) practice first on a small sample before treating a large batch of seeds.

Although the germination of properly treated seed will not be significantly affected, the storage life of the seed may be shorter.

Equipment need for this process is as follows: (1) large metal pot and several smaller containers; (2) bags for seed made of muslin, cotton, or nylon (panty hose will do); (3) an accurate thermometer from a science supply house, and (4) sieve, colander, vegetable steamer, or hardware cloth cut in the same diameter as the inside of the pot.

The procedure is as follows:

— Place the seeds loosely in a fabric bag, filling no more than half full, and then tie securely. Add a small weight, such as a bolt to help sink the seeds.

— Set up the treatment container so that the bag of seed doesn't touch the bottom of the pot. There are several ways to keep the bag off the bottom:
  — suspend the seed bag on a vegetable steamer,
  — set the bag in a colander, or
  — cut a circle of hardware cloth and suspend it at least ½" off the bottom by using small stones.

— Fill the treatment container about half full, and using an accurate thermometer, bring the water to the treatment temperature.

— Pre-soak the seeds in water, between 100 to 105°F (38 to 41°C) for five to ten minutes to prewarm the seeds.

— Once the temperature in the treatment container is constant, add the bag of seed. While treating the seed, stir the water slowly and constantly, while keeping an eye on the temperature and the clock. An accurate thermometer is essential because temperatures one or two degrees too high can injure the seed, and one or two degrees too low will fail to kill the pathogen. To help adjust the temperature, cold or hot water can be added while stirring.

- Remove the seed, and cool by dipping the bag briefly in lukewarm water, and then in cool water.
- Open the bag and spread out the seed to dry on a hard surface or fine screen. Dry at room temperature using a fan for supplemental ventilation, and stir the seeds several times daily until dry.

Germination Testing

Most growers prefer to sell seed wholesale and leave germination testing to the seed company that has the equipment and supplies necessary to do accurate testing. You can perform your own germination tests for the purpose of testing the germination of old seed lots, or testing new seed lots that you may want to store for some period of time. Tests that you do on your own may not necessarily be legal for the seed trade, but they will be accurate enough for your own use.

For supplies you will need blotters or thick paper towels, tweezers or forceps for picking up or placing seed, zip lock bags, a misting bottle, and perhaps plastic boxes used for honeycomb (for light-dependent germinators). As a substitute for a germination chamber you can place the seeds on top of a refrigerator or hot water heater. To build a simple germination chamber, obtain an old student refrigerator (approximately two cubic feet capacity) and remove the compressor.

Alternatively, an insulated plywood box can be used as a chamber. Cover any holes with caulk, duct tape, or other suitable material. For monitoring temperature, drill a hole in the side or top to insert the probe of a thermometer. A maximum-minimum thermometer is recommended because it records the temperature history and is useful for determining if the temperature has exceeded the desired range. For a heat source, use a 40-watt bulb in a porcelain fixture and wire it in series with a computer fan and dimmer switch to control the heating rate.

The dimmer shouldn't be run below 75% of capacity because the computer fan may stall or not start up when the current is switched on. The electrical supply cord from the chamber is connected to a timer to give a 16-hour warm period at 86°F (30°C) and an 8-hour cool period (ambient room temperature). To perform a simple germination test, count out 100 seeds and place them somewhat equidistant from each other, usually not closer than ½", spreading them out on the upper half of the moist blotter or paper towel.

The bottom half of the blotter is folded over the top, and then the towel is folded over with a fold about every two to three inches and held in place loosely by a rubber band. The towels or blotters are then placed vertically in plastic trays refrigerator containers, which are placed in the germination chamber or suitable location. The trays or containers should be covered loosely with zip-lock or plastic bags to keep the moisture in the towels, but they should not be completely closed because seeds need air. Towels are inspected twice a day and misted as necessary.

Most vegetable seeds germinate in about 7 to 14 days. During the first count remove the seeds that have germinated and when the final count is done, the remaining seeds are scored and the paper towel discarded. Seeds that require light are placed on blotter paper within plastic honey boxes. The light from the germination chamber will be sufficient to satisfy the light requirement. Some cool-weather crops such as lettuce should be germinated at room temperature. Most varieties of lettuce will not germinate above 80°F (27°C). Although a test of 100 seeds will give you a fairly reliable germination test, a test of 400 seeds will remove most of the statistical error and sampling error in testing. For an accurate germination test, you should contact your state seed-testing laboratory to find out what is required in terms of fees (if any) and sample size for testing. Virginia requires a minimum 1000 seeds for a test. State seed testing laboratories and seed companies follow a strict protocol for testing each type of seed.

Germination Enhancement Techniques

There are several techniques that can be used to improve the germination of seed lots. These are especially useful when it is necessary to bring a lot of seed above the Federal germination standard.

Mixing Seeds

The easiest way to enhance germination is to combine two lots of germination tested seed, one having a high germination score and the other having a lower germination score. The two lots are thoroughly mixed and the new germination percentage is calculated. For example, mixing 600 grams of 90% germination seed with 400 grams of 70% germination seed yields 1000 grams of 82% germination seed. This technique might be used by a seed company in an effort to meet market demand when the seed is in short supply or to avoid discarding seed.

If this technique is used, the seed lot with the lower germination should not be too low in vigor or germination, otherwise there will be too much variability in the stand of plants produced from such seed. This technique should not be used if one of the seed lots is significantly below Federal standard. The other concern is that organic growers should make certain that both lots are certified organic. The mixed seed is given a new lot identification number, but a record must be kept indicating the origin and germination data of the original seed lots.

Types of Seed Processing

Types of seed that have been cleaned using winnowing as part of the seed cleaning process are candidates for these methods. Seeds of pumpkins, squash, melons, cucumbers, sunflowers, lettuce, and flower seed can be aggressively re-winnowed to remove lighter, immature seed that is not well filled out. Such seed will be lighter than the good seed, which is denser and more resistant to being carried away by the air current during the cleaning process.

Microwave Treatment

When seeds are subjected to a brief exposure of microwaves they respond with enhanced germination. This treatment must be carefully controlled, is appropriate only for small seed lots, and is mainly used to help rescue low germination seed, especially seed of endangered varieties. This treatment on a lot of low-germination pepper seed, using an exposure time of 15 seconds to increase the germination by about 25%. Longer treatment times were not attempted. This method is not recommended for large seed lots because of the difficulty in obtaining equal microwave exposure in all areas of the treated seed lot.

The procedure involves placing a thin layer of seed in the centre of the microwave oven and microwaving for approximately 10 to 30 seconds depending on the size and water content of the seed. Sunflower seed, when subjected to 30 seconds of microwave treatment, germinates faster than nonmicrowaved seeds, but when microwaved for 60 seconds they fail to germinate. Microwave radiation selectively heats the water within the seed. In addition, fats and oils are heated, though to a lesser extent. Fava beans microwaved for 30 seconds show a seven-fold increase in certain nutrients.

The heating process speeds up metabolism and there is some speculation that the heating may affect the permeability of the lipid-containing membranes, thereby causing cellular changes leading to enhanced germination. If the seed is heated too rapidly or too long, certain biological molecules (proteins) such as enzymes are denatured, losing their enzymatic activity, and the seed is killed. Seeds that have a low seed moisture content may need longer exposures than seeds with a high seed moisture content. For most seed treatments the initial exposure should be anywhere from 10 to 15 seconds, with additional trials at 10 or 15 second increments.

If this procedure is used for germination enhancement, some preliminary experiments should be done to determine the optimum exposure for germination enhancement for a particular type of seed. The effect of microwaves on seed germination is not well documented, so reports of any experiments are welcome and should be forwarded to the author. Because microwaving seed constitutes a seed treatment, check with your organic certifier before planting the seed or using it in commerce. In any case, microwaved seed needs to be retested after treatment.

Other Techniques

There are a number of other techniques that can be used to enhance seed germination. These include use of seed coatings, sodium hypochlorite treatment, and in the case of dormant seeds use of scarification and stratification. These techniques are beyond the scope of this publication, but it is worth noting that certain recently harvested vegetables seeds, especially pepper seed, can exhibit dormancy or slow germination. Post-harvest dormancy in most peppers can be overcome by curing the seed for at least six weeks at a temperature of 70 to 80°F (21 to 27°C).

Labelling and Record Keeping

Labelling

During the growing season, through planting, harvest, and processing it is essential that all crops be properly labelled. This is especially important once the crop is harvested, as many seeds may look alike. Labels should be on every container and should travel through every step of process and cleaning. While this may seem obvious, it is very easy to think that you will remember what was in a particular container, but it is surprisingly easy for several hours to slip by, and then to get confused about the identity of a particular seed lot. Also, when seed is harvested at different times from the same variety, it is also useful to put a date on the label. This can be important later on if there are questions about purity due to improper isolation, disease, or other issues that arise retrospectively.

Record Keeping

You should keep a log of most everything you do with your crops, including information about planting times and location for each variety, lot number and source of seed planted, germination percent or comments about germination and vigor, number of plants transplanted, maturity data, yield data, insect problems, organic fertilisers and amendments, and other data which may become useful later. Your garden or farm log is also important for organic certification.

Detailed records should be kept in a card file or computer database regarding variety characteristics and performance. Such data should include an accession number and information about seed source, lot number, year last grown, maturity data, plant characteristics (growth habit, yield, productivity, colour and shape of fruit), disease resistance or susceptibility, and taste comparisons with other varieties. Record keeping can be time consuming, so from the perspective of time management it is important to keep in mind that the real value of information is information re-use.

Shipping Seeds

All seed must be thoroughly dry prior to shipment. Typically the minimum recommended time between harvest and shipment is six weeks. During the last three weeks of this period, the seed should be stored in a climate-controlled environment to allow the seed to cure. This means that seeds shipped in the late summer should be cured in an air-conditioned environment, and seeds shipped in winter should be stored in a heated environment prior to shipment.

Small lots of seeds can be packed in ordinary envelopes, provided that all edges of the envelope are sealed with tape, and the envelope placed inside a second envelope. The seams of the inner envelope must be sealed so that seeds do not sift out of the envelope during shipment. Zip-lock bags are available in various sizes, but it is best to use 4 mil thick bags. Zip-lock bags are fine to use, but if the bag is to hold several ounces of seed, especially heavy or sharp seed it is best to double bag these.

The zipped edge should be folded over and taped or banded to help prevent accidental opening of the zipped edge. All air should be sucked out of the bags to avoid popping. The seeds should then be packed in a cardboard box or a padded mailer. Large seeds, such as peas, beans, and corn can be packed in grocery bags, but must be triple-bagged, and taped securely to prevent breakage of the bag during shipment. There should be enough padding inside the container so that there is no rattling or shifting of contents. During the process of delivery, the packing material will become compacted. Newspaper tends to compact too much, so stiffer materials such as crumpled grocery bags are recommended.

One of the most enlightening lessons of running a mail-order business is opening a package that has been shipped across country and back. That provides great feedback about your skill in packing. Mark each seed container with your name and the variety name on a slip of paper, both inside the container and on the outside of the seed container. Mark on the outside of the shipping package in large letters "Perishable - Live Seeds Keep Cool and Dry." Include shipping address and a packing list with the package. Address labels should always be covered with clear packing tape; otherwise the package conveyor belt may snag the edge of the label and rip it off the package. All seams need to be securely taped. Sometimes packages can be left outside in the rain by the delivery person, so you don't want water entering an open seam.

References

Grabe, D. F. and D. Isely., "Seed storage in moisture-resistant packages" *Seed World,* Jan 24, 1969, p. 4.

Harrington, J.F., "Vegetable seed germination temperatures," *Agricultural Extension leaflet,* University of California, Davis, 1954.

Navazio, J., "Seed-born diseases" (lecture handout), *Organic Seed Alliance,* Port Townsend, WA, 2003.

3

Processing of Horticultural Crops

Some produce requires blanching before freezing or drying, and fruits such as apples, pears, peaches and apricots are sometimes treated with sulfur being dried. Blanching by boiling water bath or in steam ends certain enzymatic reactions in the product and helps retain colour and flavour after processing. Sulfuring (burn one tablespoon of sulfur powder per pound (35 ml per kg) of fruit or dip fruit in a 1% potassium metabisulfite solution for one minute) helps prevent darkening, loss of flavour

A low cost sulfuring box can be constructed from a large cardboard box that is slashed in several places to allow adequate ventilation. Trays for drying be stacked using bricks and wooden spools as spacers. The trays must be made completely of wood, since sulfur fumes will corrode metal The entire assembly must be located out of doors, preferably on bare soil. Use one tablespoon of sulfur powder per pound (35 mls per kg) of fruit. Place the sulfur in a container well away from the side of the box since it will become quite hot. Seal the bottom edges of the box with soil.

Solar Drying

Horticultural produce can be dried using direct or indirect solar radiation. The simplest method for solar drying is to lay produce directly upon a Hat black surface and allow the sun and wind to dry the crop. Nuts can be dried effectively in this way. Simple direct driers can be made from trays of screening material propped upon wooden or concrete blocks to allow air to circulate under the produce.

A layer of cheesecloth can be draped loosely over the produce, protecting it from insects and birds while drying. A simple method for solar drying is to construct a raised platform from wood and cover the frame with loosely woven mats. In the illustration below, sliced fresh tomatoes are being dried in direct sunlight on straw mats. Air can pass over and below the produce, speeding drying and reducing losses due to overheating.

Straw Mats

In order to improve the efficiency of drying, some sort of structure must be used to capture solar radiation. Various types of solar driers have been developed and are illustrated below. More complex models of solar driers have glass or clear plastic windows that cover the produce, providing some protection from insects while capturing more of the heat of the sun.

Direct Solar Drier

Indirect driers are constructed so the sun shines upon a solar collector (a shallow box, the insides painted black, topped with a pane of glass) heating air which then moves upward through a stack of four to six trays loaded with produce.

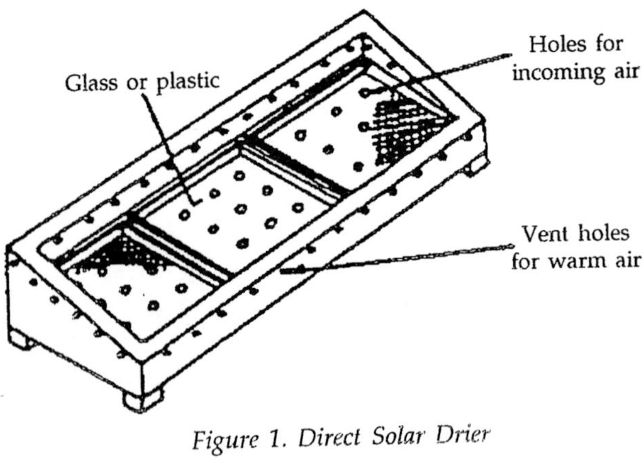

Figure 1. Direct Solar Drier

The solar drier for cassava chips illustrated below consists of a solar collector, a fan, and drying bin. The solar collector is constructed on a concrete base, using a layer of fine stones and two layers of concrete blocks, all covered with polyethylene. The air heated within the collector is then forced through the perforated floor of the drying bin. The walls at the top of the bin, below the overhanging roof, are made of screen, allowing for easy movement of air through the produce.

Forced-air Dehydrators

Nut crops can be dried in bulk using a dehydrator that combines a steady stream of air with an external source of heat. The plenum chamber below the produce is covered with a floor of perforated sheet metal or wooden slats. A fan located between the furnace and the plenum chamber moves the hot air through the drying produce.

Oil-burning Dehydrators

The batch-dryer illustrated below is constructed of wood, has an axial type fan and burns kerosene or diesel oil. A wide variety of dryers are available from manufacturers around the world.

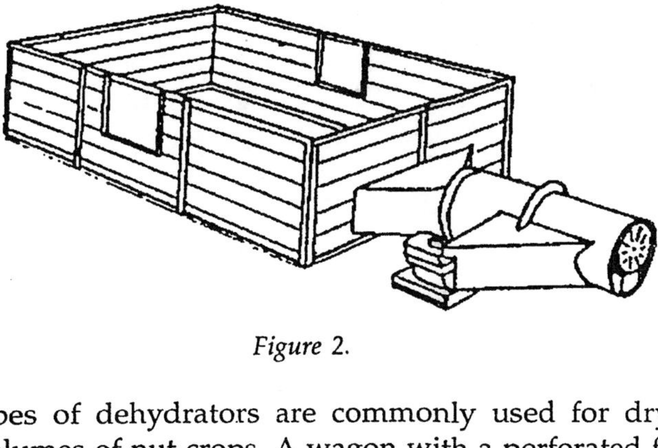

Figure 2.

Two types of dehydrators are commonly used for drying small volumes of nut crops. A wagon with a perforated floor can be transported from the field and connected to a portable burner batch drying. A stationary "pot-hole" dehydrator is designed to move heated air along a plenum under a fixed platform Individual bins of nuts are placed upon the platform and are dried as heat rises up through the perforated floor.

Electric Dehydrators

A simple electric dehydrator can be constructed using plywood, sheet metal a small fan, five household lightbulbs with porcelain mounting fixtures and some screening material. The drier shown below is 32 inches long by 21 inches wide and 30 inches high, and contains racks for five trays. The fan and the sheet metal lining the bottom compartment help conduct heat upward through the box.

Oven Drying

Fruits and vegetables can be dried in a home oven if the oven can be run at a low temperature. Place the prepared produce on baking or metal screen trays, set the oven temperature at 140 degrees F and leave the door ajar (2 to 4 inches). Drying time can be reduced if ventilation is increased by using a small fan placed outside the oven.

Figure 3.

Drying Flowers

Cut flowers can be air dried by hanging upside down or while supported by chicken wire. Certain flowers will look more natural if left standing in a vase while they dry. Anthurium dries best if left to dry very slowly. Cut the stems at a sharp angle, and place the flowers into a vase containing two inches of water. In all cases, flowers should be left to air dry in a dark, well area.

— Flowers that dry best if left standing: strawflower, delphinium, larkspur, okra pods

— Flowers that dry best while hanging upside down: chysanthemum, amaranthus, African daisy, statice, marigold

Cut flowers can be dried quickly and easily in sand or silica gel. Sand used for drying flowers should be clean? smooth and the finer the better. Starting with one inch of sand in a container, place the flower to be dried on the sand and gently cover the entire flower with more sand. The container should be left uncovered and flowers should be dried in about three weeks. Flowers that dry well in sand are shasta daisy, lily-of-the-valley, cosmos, dahlia, sweet william carnation, stock, freesia and narcissus.

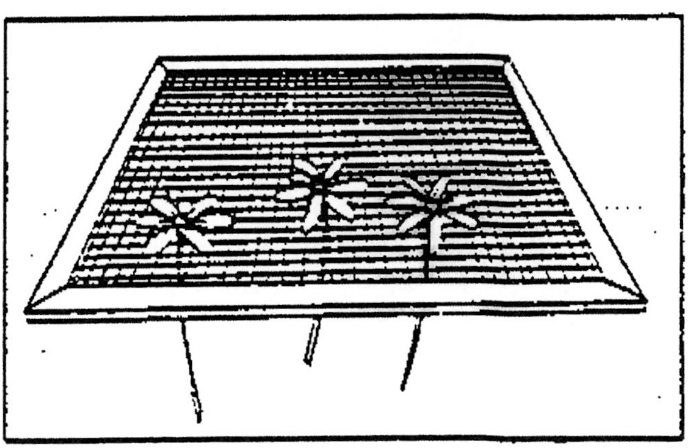

Figure 4. African daisy supported on a screen of chicken wire:

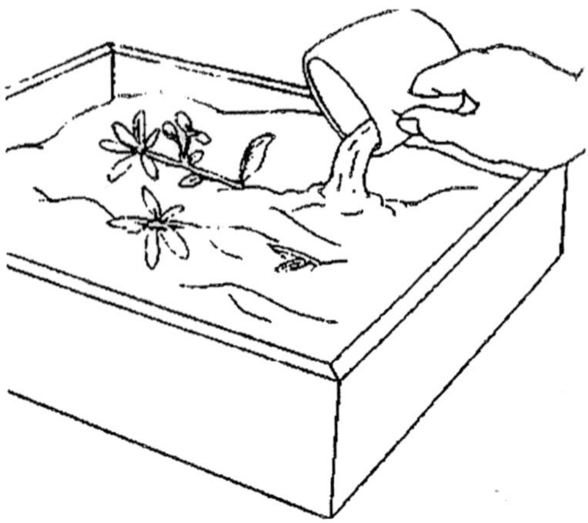

Figure 5. Drying flowers in sand:

Silica gel is relatively expensive but reusable if heated to dry out the gel between uses. To use, cover the flower as with sand, then tightly seal the container. Check for drying in two to three days. Silica gel is especially useful for drying fragile plants and flowers with delicate colours. Flowers that dry best in silica gel are allium, anemone, cornflower, roses, tulip and zinnia.

Extraction of Oils from Aromatic Plants

The steam extraction unit illustrated below was first constructed for experimental extraction of essential oils from small quantities of aromatic plants. This 500 liter (130 gallon), stainless steel vessel may be used by small scale processors, as it extracts up to 100 ml (3.5 fluid ounces) of essential oils per distillation and can be operated by a single individual.

The model shown below is portable if mounted on a trailer. Steam is introduced at the bottom of the vessel, and moves through a layer of plant material which rests upon a perforated stainless steel plate. Steam, water vapour and extracted volatiles exit the tank at the top, then pass thorough a water cooled aluminum condenser. The tank can be tilted for ease m emptying, cleaning, and reloading.

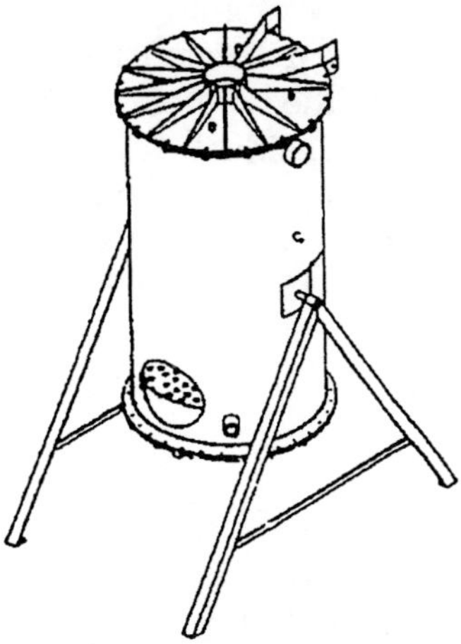

Figure 6. The steam extractor:

Figure 7. The condenser:

Canning

Two types of canners are commonly used to process horticultural crops. The first is a water bath canner, which is a large pot with a loose cover and a rack to hold jars off the bottom The pot should be deep enough to cover the canning jars by one to two inches and still have another inch of space to allow brisk boiling. The diameter of the pot should be no more than four inches wider than the diameter of the stove's burner to ensure even heating.

Acidic foods such as fruits, tomatoes, pickles and relishes, and high sugar foods such as jams, jellies, syrups and marmalades can be safely processed using a boiling water bath. A pressure canner is recommended for processing low acid foods such as vegetables. A pressure canner is a specially made heavy pot with a locking lid, an inner rack and a steam vent in the kid. The vent can be adjusted using a weight, value or screw, depending on the type of canner. A pressure gauge registers the air pressure inside the canner. A dial gauge gives a reading of the actual pressure, while weighted gauges will rock gently when the canner is at the proper pressure. Ten pounds of pressure at 115 C (240 F) is recommended for canning vegetables.

— Pressure Canners

— Dial Gauge

— Weighted Gauges

There are three types of glass canning jars used for processing horticultural crops. The ball type jar and the zinc capped jar both require rubber rings as seals. These can sometimes be difficult to obtain, but if locally available, make excellent containers. Currently the canning jar with a two-pieced lid is the most commonly used container. No matter which jar is used, when filling containers, it is important to leave a small amount of headspace to allow for expansion of the food while processing. If a jar is filled too full, it may explode. If too much headspace is left, the food may spoil, since all the extra air may not be driven out during processing.

Juicing

Fruits

To process tomatoes or fruits to juices, fruits are simmered in water or their own juice in a stainless steel, glass or enamelware pot. When tender, the product is cut into pieces and pressed through a food mill, colander or several layers of cheesecloth. Sugar or lemon juice can be added, to taste.

The juices must then be either frozen or canned for storage. Juices can be frozen in jars or freezer containers (leave 1/2 inch headspace). Most fruit juices can be canned in a boiling water bath for 20 minutes, but apple and grape juices can be processed in hot water (82 C or 180 F) for 30 minutes.

Vegetables

Vegetables should be chopped or shredded, then simmered for 45 to 50 minutes until mushy. The juice can then be pressed or strained from the vegetable pulp, and frozen or canned. Canning vegetable juices requires processing at ten pounds of pressure in a pressure canner. Pints should be processed for 55 minutes, and quart jars for 85 minutes.

Other Methods of Processing

Freezing

Most vegetables should be blanched before freezing to prevent loss of flavour and colour during storage. Freezing temperatures are best set between -21 to -18 C (0 to 5 F). Packages for freezing should be moisture proof and vapour proof and contain as little air as possible to prevent oxidation during storage. Heavy plastic bags, heavy aluminum foil, glass freezer jars and waxed freezer cartons all make good containers.

Jellies, Jams and Preserves

Making jams, jellies and other high sugar preserves requires a balance of fruit, acid, pectin and sugar for best results. Underripe fruits contain more pectin than ripe fruits, and apple juice is a good source of natural pectin. If fruits are low in acid, lemon juice can be added to the mixture of fruit and sugar. Cane or beet sugar is better for making preserves than honey or corn syrup.

To preserve fruits, cook on medium heat until the mixture "sheets" from a spoon. Avoid overcooking since this will lower the jelling capacity of the mixture. Pour into containers and seal with paraffin wax (jellies only). The other preserves should be processed in a boiling water bath for five minutes.

Fermentation

When lactic acid bacteria in foods convert carbohydrates to lactic acid, food is preserved by the resulting low pH.

Sauerkraut (cabbage) and wine (grapes) are two examples of thousands of fermented foods made around the world.

Acidification

Pickling is a simple processing method that can be used with many types of fruits and vegetables. Brine solution (9 parts cider or white vinegar, 1 part non-iodised salt, 9 parts water, plus flavorings and spices) is poured over the product into glass canning jars (leave 1/2 inch headspace). Brined pickles are sealed and left at ambient temperature for three or more weeks, while fresh pack pickles are processed in a boiling water bath for 10 minutes.

References

Campbell-Platt, G., *Fermented Foods of the World: Dictionary and Guide*, Stonam, Massachussetts: Butterworth Heineman, 1987.

Fellows, P. and Hampton, *A Small-Scale Food Processing A Guide to Appropriate Equipment*, London: Intermediate Technology Publications, 1992.

Hunsigi, G., *Agricultural Fibres for Paper Pulp*, Outlook on Agriculture 18 (3): 96-103, 1989.

Talbot, M. T. and Fletcher, J.H., *Design and Development of a Portable Forced-Air Cooler*, Proceedings of the Florida State Horticultural Society 106:249-255, 1993.

4

Fruit Specific Processing Technologies

There is a distinction between maturity and ripeness of a fruit. Maturity is the condition when the fruit is ready to eat or if picked will become ready to eat after further ripening. Ripeness is that optimum condition when colour, flavour and texture have developed to their peak. Some fruit is picked when it are mature but not yet ripe. This is especially true of very soft fruit like cherries and peaches, which when fully ripe are so soft as to be damaged by the act of picking itself.

Further, since many types of fruit continue to ripen off the tree, unless they were to be processed quickly, some would become overripe before they could be utilised if picked at peak ripeness. From a technological point of view, fruit characterisation by species and varieties is performed on the basis of physical as well chemical properties: shape, size, texture, flavour, colour/pigmentation, dry matter content (soluble solids content), pectic substances, acidity, vitamins, etc. These properties are directly correlated with fruit utilisation.

The proper time to pick fruit depends upon several factors; these include variety, location, weather, ease of removal from the tree (which change with time), and purpose to which the fruit will be put. For example, oranges change with respect to both sugar and acid as they ripen on the tree; sugar increases and acid decreases. The ratio of sugar to acid determines the taste and acceptability of the fruit and the juice.

For this reasons, in some countries there are laws that prohibit picking until a certain sugar-acid ratio has been reached. In the case of much fruit to be canned, on the other hand, fruit is picked before it is fully ripe for eating since canning will further soften the fruit.

Many quality measurements can be made before a fruit crop is picked in order to determine if proper maturity or degree of ripeness has developed. Colour may be measured with instruments or by comparing the colour of fruit on the tree with standard picture charts. Texture may be measured by compression by hand or by simple type of plungers. As fruit mature on the tree its concentration of juice solids, which are mostly sugars, changes. The concentration of soluble solids in the juice can be estimated with a refractometer or a hydrometer.

The refractometer measures the ability of a solution to bend or refract a light beam which is proportional to the solution's concentration. A hydrometer is a weighted spindle with a graduated neck which floats in the juice at a height related to the juice density. The acid content of fruit changes with maturity and affects flavour. Acid concentration can be measured by a simple chemical titration on the fruit juice. But for many fruits the tartness and flavour are really affected by the ratio of sugar to acid.

Percentage of soluble solids, which are largely sugars, is generally expressed in degrees Brix, which relates specific gravity of a solution to an equivalent concentration of pure sucrose. In describing the taste of tartness of several fruits and fruit juices, the term "sugar to acid ratio" or "Brix to acid ratio" are commonly used. The higher the Brix the greater the sugar concentration in the juice; the higher the "Brix to acid ratio" the sweeter and lees tart is the juice.

Harvesting and Preprocessing

The above and other measurements, plus experience, indicate when fruit is ready for harvesting and subsequent processing. A large amount of the harvesting of most fruit crops is still done by hand; this labour may represent about half of the cost of growing the fruit. Therefore, mechanical harvesting is currently one of the most active fields of research for the agricultural engineer, but also requires geneticists to breed fruit of nearly equal size, that matures uniformly and that is resistant to mechanical damage.

A correct manual harvesting includes some simple but essential rules:

— the fruit should be picked by hand and placed carefully in the harvesting basket; all future handling has to be performed carefully in order to avoid any mechanical damage;

— the harvesting basket and the hands of the harvester should be clean;

— the fruit should be picked when it is ready to be able to be processed into a quality product depending on the treatment which it will undergo.

It is worth emphasising the fact that the proximity of the processing centre to the source of supply for fresh raw materials presents major advantages; some are as follows:

— possibility to pick at the best suitable moment;

— reduction of losses by handling/transportation;

— minimises raw material transport costs;

— possibility to use simpler/cheaper receptacles for raw material transport.

Once it has left the tree, the organoleptic properties, nutritional value, safety and aesthetic appeal of the fruit deteriorates in varying degrees. The major causes of deterioration include the following:

— growth and activity of micro-organisms;

— activities of the natural food enzymes;

— insects, parasites and rodents;

— temperature, both heat and cold;

— moisture and dryness;

— air and in particular oxygen;

— light and

— time.

The rapidity with which foods spoil if proper measures are not taken is indicated in Table 1.

Table 1. Useful storage life of some food products

Food Products (Days) at 21°C (70°F)	Generalised Storage Life
Animal Flesh, Fish, Poultry	1-2
Dried, salted, smoked meat and fish	360 and more
Fruits	1-7
Dried Fruits	360 and more
Leafy Vegetables	1-2
Root Crops	7-20

Fruit Reception

Fruit reception at the processing centre is performed mainly for following purposes:

— checking of sanitary and freshness status;

— control of varieties and fruit wholeness;

— evaluation maturity degree;

— collection of data about quantities received in connection to the source of supply: outside growers/ farmers, own farm.

Variety control is needed in order to identify that the fruit belongs to an accepted variety as not all are suitable for different technological processes. Fruit maturity degree is significant as industrial maturity is required for some processing/preservation methods while for others there is the need for an edible maturity when the fruit has full taste and flavour. Special attention is given to size, appearance and uniformity of fruit to be processed, mainly in the form of fruit preserved with sugar using whole/half fruits ("with fruit pieces").

Some laboratory control is also needed, even if it not easy to precisely establish the technological qualities of fruit because of the absence of enough reliable rapid analytical methods able to show eventual deterioration. The only reliable method for evaluating the quality is the combination of data obtained through organoleptic/taste controls and by simple analytical checks which are possible to perform in a small laboratory: percentage of soluble solids by refractometer, consistency/texture measured with simple penetrometers, etc.

Temporary Storage Before Processing

This step has to be as short as possible in order to avoid flavour losses, texture modification, weight losses and other deterioration that can take place over this period.

Some basic rules for this step are as follows:

— keep products in the shade, without any possible direct contact with sunlight;

— avoid dust as much as possible;

— avoid excessive heat;

— avoid any possible contamination;

— store in a place protected from possible attack by rodents, insects, etc.

Cold storage is always highly preferred to ambient temperature. For this reason a very good manufacturing practice is to use a cool room for each processing centre; this is very useful for small and medium processing units as well. The type of analysis for audits will be adapted to the specific fruits and vegetables that are received/ processed. An excellent indication of a good temporary storage is the limited weight loss before processing, which has to be below 1.0%-1.2%.

Washing Efficiency

Harvested fruit is washed to remove soil, micro-organisms and pesticide residues. Fruit washing is a mandatory processing step; it would be wise to eliminate spoiled fruit before washing in order to avoid the pollution of washing tools and/or equipment and the contamination of fruit during washing.

Washing efficiency can be gauged by the total number of microorganisms present on fruit surface before and after washing—best result are when there is a six fold reduction. The water from the final wash should be free from moulds and yeast; a small quantity of bacteria is acceptable. Fruit washing can be carried out by immersion, by spray/ showers or by combination of these two processes which is generally the best solution: pre-washing and washing.

Some usual practices in fruit washing are:

— addition of detergents or 1.5% HCl solution in washing water to remove traces of insect-fungicides;

— use of warm water (about 50°C) in the pre-washing phase;

— higher water pressure in spray/shower washers.

Washing must be done before the fruit is cut in order to avoid losing high nutritive value soluble substances (vitamins, minerals, sugars, etc.).

Fruit Sorting

Fruit sorting covers two main separate processing operations:

— removal of damaged fruit and any foreign bodies (which might have been left behind after washing);

— qualitative sorting based on organoleptic criteria and maturity stage.

Mechanical sorting for size is usually not done at the preliminary stage. The most important initial sorting is for variety and maturity. However, for some fruit and in special processing technologies it is advisable to proceed to a manual dimensional sorting (grading).

Trimming and Peeling of Fruit

This processing step aims at removing the parts of the fruit which are either not edible or difficult to digest especially the skin. Up to now the industrial peeling of fruit and vegetables was performed by three procedures:

— mechanically;

— by using water steam;

— chemically; this method consists in treating fruit and vegetables by dipping them in a caustic soda solution at a temperature of 90 to 100° C; the concentration of this solution as well as the dipping or immersion time varying according to each specific case.

Cutting

This step is performed according to the specific requirements of the fruit processing technology.

Heat Blanching

Fruit is not usually heat blanched because of the damage from the heat and the associated sogginess and juice loss after thawing. Instead, chemicals are commonly used without heat to inactivate the oxidative enzymes or to act as antioxidants and they are combined with other treatments.

Ascorbic Acid Dip

Ascorbic acid or vitamin C minimises fruit oxidation primarily by acting as an antioxidant and itself becoming oxidised in preference to catechol-tannin compounds. Ascorbic acid is frequently used by being dissolved in water, sugar syrup or in citric acid solutions. It has been found that increased acidity also helps retard oxidative colour changes and so ascorbic acid plus citric acid may be used together. Citric acid further reacts with (chelates) metal ions thus removing these catalysts of oxidation from the system.

Sulphur Dioxide Treatment

Sulphur dioxide may function in several ways:

— sulphur dioxide is an enzyme poison against common oxidising enzymes;

- it also has antioxidant properties; i.e., it is an oxygen acceptor (as is ascorbic acid);
- further SO_2 minimises non enzymatic Maillard type browning by reacting with aldehyde groups of sugars so that they are no longer free to combine with amino acids;
- sulphur dioxide also interferes with microbial growth.

In many fruit processing pre-treatments two factors must be considered:

- sulphur dioxide must be given time to penetrate the fruit tissues;
- SO_2 must not be used in excess because it has a characteristic unpleasant taste and odour, and international food laws limit the SO_2 content of fruit products, especially of those which are consumer oriented.

Commonly a 0.25 % solution (except for semi-processed fruit products which are industry oriented and use a 6% solution) of SO_2 or its SO_2 equivalent in the form of solutions of sodium sulphite, sodium bisulphite or sodium/potassium metabisulphite are used. Fruit slices are dipped in the solution for about two to three minutes and then removed so as not to absorb too much SO_2. Then the slices are allowed to stand for about one to two hours so that the SO_2 may penetrate throughout the tissues before processing.

Sulphur dioxide is also used in fruit juice production to minimise oxidative changes where relatively low heat treatment is employed so as not to damage delicate juice flavour. Dry sulphuring is the technological step where fruit is exposed to fumes of SO_2 from burning sulphur or from compressed gas cylinders; this treatment could be used in the preparation of fruits (and some vegetables) prior to drying / dehydration.

Sugar Syrup

Sugar syrup addition is one of the oldest methods of minimising oxidation. It was used long before the causative reactions were understood and remains today a common practice for this purpose. Sugar syrup minimises oxidation by coating the fruit and thereby preventing contact with atmospheric oxygen. Sugar syrup also offers some protection against loss of volatile fruit esters and it contributes sweet taste to otherwise tart fruits. It is common today to dissolve ascorbic acid and citric acid in the sugar syrup for added effect or to include sugar syrup after an SO_2 treatment.

Fresh Fruit Storage

Some fruit species and specially apples and pears can be stored in fresh state during cold season in some countries' climatic conditions. Fruit for fresh storage have to be autumn or winter varieties and be harvested before they are fully mature. This fruit also has to be sound and without any bruising; control and sorting by quality are mandatory operations.

Sorting has to be carried out according to size and weight and also by appearance; fruit which is not up to standard for storage will be used for semi-processed product manufacturing which will be submitted further to industrial processing. Harvested fruit has to be transported as soon as possible to storage areas. Leaving fruit in bulk in order to generate transpiration is a bad practice as this reduces storage time and accelerates maturation processes during storage.

Fruit Drying and Dehydration Technology

For fruit with a high sugar content drying temperatures have to be lower at initial stage and then increase to the maximum acceptable; for fruit with lower sugar level the temperatures are applied in a reverse order.

Processing of Fruit Bars

The fruit bar processing method developed for FAO only involves a single major operation, which it drying the fruit pulp after mixing it with suitable ingredients. It can be used to produce mango, banana, guava or mixed fruit bars. A dual-powered dryer, working by solar energy during the day and by electric or steam power at night and on rainy days, with cross-flow movement of air and controlled temperature (from 55° C at the beginning of processing to a high of 70° C), is well suited for dehydration of the pulp to the desired moisture level of 15 to 20%.

Mango fruit bar: Fully ripe mangoes are selected and washed in water at room temperature. The peeled fruit is cut into slices and passed through a helicoidal pulper to extract the pulp. The required amount of sugar to adjust the Brix (the unit measure for total solids in fruits) of the mixed pulp to 25 degrees Brix is then added. Two grams of citric acid per kilogram of pulp (or 20 ml of lime or lemon juice) are added to inhibit possible growth of microorganisms during drying. The mixture is then heated for two minutes at 80° C and partially cooled; the heat treatment serves to inactivate the enzymes and destroy the microorganisms.

Potassium or sodium metabisulphite is added (two grams per kg of prepared mixture), so that the concentration of SO_2 is 1000 ppm. The mixture is then transferred to stainless steel trays which have been previously smeared with glycerine (40 ml/m^2). Each tray must be loaded with 12.5 kg of mixture per square metre.

Drying could be carried out by a dual-powered dryer for a total of 26 hours:

— 10 hours by solar energy at about 55° C and

— 16 hours by electric or steam power at 70° C.

At the end of the drying operation, when moisture content is between 15 and 20%, the pieces of suitable shape and size are wrapped in cellophane paper, packed in cartons and stored at ambient air temperature. Pieces of unsuitable shape and size are further cut into small pieces and used to prepare, along with peanuts and cashews, a variety of "cocktail mixtures".

Banana fruit bar: Banana varieties which give smooth pulp without serum separation must be used for this purpose. Ripe, suitable fruit is selected. The hand-peeled fruits are soaked in

0.3 per cent citric acid solution for about 10 minutes (lime or lemon juice can replace citric acid). The drained fruit are pulped to obtain smooth pulp.

Guava fruit bar: A mixture of pink and yellow varieties is best suited for preparing the bar. The washed fruit is hand peeled and stem and blossom ends trimmed. The peeled fruit is cut into quarters which are passed through a helicoidal extractor to separate seeds and fibrous pieces (the holes in the stainless steel screen should be between 0.8 and 1.10 mm). To get the maximum yield of pulp, the material is passed through the extractor twice. After adjusting the refractometric solids to 25 degrees Brix, the fruit bar can be prepared by following the same procedure as for mango pulp.

Mixed fruit bar: Mango and banana pulp, as well as papaya and banana pulp, can be mixed in a calculated ratio for preparing mixed fruit bars. The rest of the procedure is the same as in the case of pure mango pulp.

Packing and storage: The dried pulp is removed from the dryer and cut into square pieces of 5 x 5 cm at a thickness of about 0.3 cm. These pieces, arranged in three layers make up blocks of about 0.9 cm thickness weighing between 25 and 28 grams. An unit pack consist of two such blocks and weights between 50 and 56 grams. Each block is separately wrapped in cellophane and the unit pack is filled in a printed cellophane bag of size 15 x 8 cm. Two hundred unit packs are packed in a master carton of size 34 x 22 x 14 cm, with a net weight of about 10 kg. Shelf-life is about one year at room temperature.

Fruit leathers: Fruit leathers are manufactured by drying/ dehydration of fruit purées into leathery sheets. The leathers are eaten as confections or cooked as a sauce. They are made from a wide variety of fruits, the more common being apple, apricot, banana, cherry, blackcurrant, grape, peach, pear, pineapple, plum, raspberry, strawberry, kiwi fruit, mango and papaya.

A description of procedures for mango, banana, guava and mixed fruit bars is given in this document. Another product with good potential is ciku leather; ciku fruit is grown in Malaysia. A standard process is carried out using ripe fruits which are washed, peeled, diced and the seeds removed. The fruits are blanched for 1 minute at 80° C and blended into puree in a food processor. Ciku leather is prepared by mixing ciku puree with 10% sugar, 10% pre-gelatinous rice flour, 150 ppm sorbic acid an 500 ppm sodium metabisulphite ($Na_2H_2SO_4$).

The mixture is cooked on a water bath at 60° C and then made into sheets 1.8 mm thick on trays spread with glycerol to reduce stickiness. This is then further dried in a forced-air dehydrator at 45° C for 3.5 hr or until the surface no longer feels sticky when touched with the fingers. The dried and cooled leathers are cut into 12 x 12 cm squares and wrapped in polypropylene (PP) of 0.1 mm thickness.

Osmotic Dehydration

Osmotic dehydration is a useful technique for the concentration of fruit and vegetables, realised by placing the solid food, whole or in pieces, in sugars or salts aqueous solutions of high osmotic pressure. It gives rise to at least two major simultaneous counter-current flows: a significant water flow out of the food into the solution and a transfer of solute from the solution into the food.

Process Variables

Main process variables are:

- — pre-treatments;
- — temperature;
- — nature and concentration of the dehydration solutions;
- — agitation;
- — additives.

In the light of the published literature, some general rules can be noted:

- — water loss and solid gain are mainly controlled by the raw material characteristics and are certainly influenced by the possible pre-treatments;

— it is usually not worthwhile to use osmotic dehydration for more than a 50% weight reduction because of the decrease in the osmosis rate over time. Water loss mainly occurs during the first 2 hr and the maximum solid gain within 30 min.;

— the rate of mass exchanges increases with temperature but above 45 ° C enzymatic browning and flavour deterioration begin to take place. High temperatures, i.e. over 60° C, modify the tissue characteristics so favouring impregnation phenomena and thus the solid gain;

— the best processing temperature depends on the food; mass exchanges are favoured by using high concentration solutions;

— phenomena which modify the tissue permeability, such as over-ripeness, pre-treatments with chemicals (SO_2), blanching or freezing, favour the solid gain compared to water loss because impregnation phenomena are enhanced;

— the kind of sugars utilised as osmotic substances strongly affects the kinetics of water removal, the solid gain and the equilibrium water content. Low molar mass saccharides (glucose, fructose, sorbitol, etc.) favour the sugar uptake;

— addition of NaCl to osmotic solutions increases the driving force for drying.

Synergistic effects between sugar and salt have also been observed.

Applications

The effects of osmotic dehydration as a pre-treatment are mainly related to the improvement of some nutritional, organoleptic and functional properties of the product. As osmotic dehydration is effective at ambient temperature, heat damage to colour and flavour is minimised and the high concentration of the sugar surrounding fruit and vegetable pieces prevents discoloration.

Furthermore, through the selective enrichment in soluble solids high quality fruit and vegetables are obtained with functional properties "compatible" with different food systems. These effects are obtained with a reduced energy input over traditional drying process. The main energy-consuming step is the reconstitution of the diluted osmotic solution that could be obtained by concentration or by addition of sugar.

Drying

Air drying following osmotic dipping is commonly used in tropical countries for the production of so-called "semi-candied" dried fruits. The sugar uptake, owing to the protective action of the saccharides, limits or avoids the use of SO_2 and increases the stability of pigments during processing and subsequent storage period. The organoleptic qualities of the end product could also be improved because some of the acids are removed from the fruit during the osmotic bath, so a blander and sweeter product than ordinary dried fruits is obtained. Owing to weight and volume reduction, loading of the dryer can be increased 2-3 times.

The combination of osmosis with solar drying has been put forward, mainly for tropical fruit. A 24 hour cycle has been suggested combining osmodehydration, performed during the night, with solar drying during the day. Two-three-fold increase in the throughput of typical solar dryers is feasible, while enhancing the nutritional and organoleptic quality of the fruits. A two-step drying process, OSMOVAC, for producing low moisture fruit products was described.

The osmotic step is performed with sucrose syrup 65-75 Brix until the weight reduction reaches 30-50%. By osmotic dehydration followed by vacuum drying puffy products with a crisp, honeycomb-like texture can be obtained at a cost comparatively lower than freeze-drying. Commercial feasibility of the process on bananas has been studied, based on the results of a semi-pilot scale operation. Osmotically dried bananas retained more puffiness and a crisper texture than simple vacuum dried ones, and the flavour lasted longer at ambient temperature. The combination of osmotic dehydration with freeze-drying has been proposed only at laboratory scale.

Appertisation

A combination of osmotic dehydration with appertisation has been proposed to improve canned fruit preserves. The feasibility of a process, called osmo-appertisation, to obtain high quality fruit in syrup, has been assessed on a pilot scale. The key point of this technique is the pre-concentration of the fruit to about 20-40 Brix, that causes, together with the enhancement of the natural flavour, an increase of the resistance of the fruit to the following heat treatment, especially for colour and texture stability.

The products obtained are stable up to 12 months at ambient temperature and show a higher organoleptic quality than canned preserved alternatives. Furthermore, because of their higher specific weight and diminished volume, the filling capacity of jars or pouches is increased.

Freezing

The frozen fruit and vegetable industry uses much energy in order to freeze the large quantity of water present in fresh products. A reduction in moisture content of the material reduces refrigeration load during freezing. Other advantages of partially concentrating fruits and vegetables prior to freezing include savings in packaging and distribution costs and achieving higher product quality because of the marked reduction of structural collapse and dripping during thawing.

The products obtained are termed "dehydro-frozen" and the concentration step is generally carried out through conventional air drying, the additional cost of which has to be taken into account. Osmotic dehydration could be used instead of air drying to obtain an energy saving or a quality improvement especially for fruit and vegetable sensitive to air drying.

Juice Extraction

An osmotic pre-step before juice extraction was reported to give highly aromatic fruit or vegetable juice concentrates.

Further Developments

So far only applications on a pilot plant scale are reported in the literature. For further developments on a larger scale, theoretical and practical problems should be solved. The industrial application of the process faces engineering problems related to the movement of great volumes of concentrated sugar solutions and to equipment for continuous operations. The use of highly concentrated sugar solutions creates two major problems. The syrup's viscosity is so great that agitation is necessary in order to decrease the resistance to the mass transfer on the solution side. The difference in density between the solution and fruit and vegetables, makes the product float. Another important aspect, so far not investigated, is the microbiological safety of the process, which should be studied thoroughly before further industrial development.

Osmoappertisation

In order to obtain an alternative to the canned fruit preserves and to maintain a high quality of the fruits, a research has been carried out on the osmoappertisation of apricots, a "combined" technique that consists in the appertisation of the osmodehydrated apricots. This technique could contributes also to the reduction of energy consumption, limits the cost of production and combines "convenience" (ready-to-eat, medium shelf-life) with many market outlets (retail, catering, bakery, confectionery, semi-finished products). Osmoappertisation combines two unit operations: dehydration by osmosis and appertisation (packaging + pasteurization).

Apricot Processing

Fresh Apricot Purée

After washing, cutting and removal of stones, apricot halves are dipped in 2% solution of sodium or potassium metabisulphite for 10 minutes. After draining, the resulting material is passed through a 0.045-in. screen pulper—finisher to produce a fresh apricot puree. The fresh apricot puree obtained in this way could be further processed in different semiprocessed (i.e. chemically or otherwise preserved products) or finished fruit products (fruit leathers, fruit bars, jams, etc.).

Concentrated Apricot Pulp

Fresh apricot halves could also be steam blanched for 5 min., passed through a 0.045-in pulper—finisher and transformed in a purée with about 14 Brix depending on the fruit quality. This purée may be concentrated in steam jacketed kettles up to 20 Brix or in other adequate equipment (e.g. a stirred vacuum evaporator) up to 28° Brix. As for fresh apricot purée, the concentrate may be further processed in various semiprocessed or finished fruit products as mentioned above and as will be described below.

Dried Apricot Leather

— From fresh fruit purée by drum drying. The fresh apricot purée at about 14 Brix could be dried to 12% moisture apricot sheets, using a double-drum dryer operating at 132 degrees C with a drum clearance of 0.008 in and speed of 45 sec per revolution.

— From fruit concentrate by drum drying. The concentrate could also be dried to 12% moisture fruit sheets by the same process as described above.

— From fresh fruit purée or from apricot concentrate by sun/solar drying or by dehydration.

Trays: For sun / solar drying or dehydration of fruit pulp, the trays must have a solid base in order to retain the liquid contents. They may be made of metal, timber or plastic. Stainless steel or plastic trays are most suitable because they are unaffected by acid fruit pulp; they are, however, expensive. A metal tray could be 75 x 50 cm in size and with side 5 cm high. The trays must keep level during drying; if the tray is not level the pulp will run to the lowest point, giving a layer of irregular depth which will dry unevenly.

Any tray which is not made of stainless steel or plastic must be covered inside with a sheet of heavy gauge plastic film to protect the pulp from chemical or bacteriological contamination. Standard sun/solar trays as described can be used by covering them inside with a sheet of plastic film to create a solid base.

Preparation before drying/dehydration: Fresh apricot purée can be directly used for next processing steps. Fruit concentrate needs to be added to potassium metabisulphite to obtain a 0.3% concentration of SO_2 in the material.

Drying/dehydration: The apricot/fruit purée or concentrate is poured into the trays to a depth of about 1.5 cm. When stainless steel or plastic trays are used they should be coated with a thin layer of glycerine to prevent sticking. The pulp is then sun/solar dried or tunnel/cabinet dehydrated; moisture content in the dried product should not exceed 14% and the SO_2 content should not be less than 1500 pp.

The dried product is wrapped in cellophane to prevent sticking, then put inside polythene bags and stored at best in tight fitting tins and sealed to prevent moisture transfer. From fresh fruit purée or from apricot concentrate, with sugar addition, and then processed by sun / solar drying or by dehydration. In some countries preference is for finished products with added sugar; and this is also interesting from a point of view of energy consumption (concentration is partially achieved by sugar dry matter) and of shelf life. The overall content in SO_2 could also be reduced as sugar is a preservation agent, the product will be close to a fruit "paste".

Reconstitution Test

In reconstitution water is added to the product which is restored to a condition similar to that when it was fresh. This enables the food product to be cooked as if the person was using fresh fruit or vegetable. All vegetables are cooked but many of the dried fruits can be used for eating after they have been soaked in water. The following reconstitution test is used to find out the quality of the dried product.

— Weigh out a sample of 35 grams from the bulked and packed final product of the previous day's production.
— Put the sample into a small container (beaker) and add 275 ml of cold water (and 3.5 g salt).
— Cover the container (with a watch-glass) and bring the water to the boil.
— Boil Gently for 30 minutes.
— Turn out the sample onto a white dish.
— At least two people should then examine the sample for palatability, toughness, flavour and presence or absence of bad flavours. The testers should record their results independently.
— The liquid left in the container should be examined for traces of sand/soil and other foreign matter.

This test can be used also to examine dried products after they have been stored for some time. Evaluation of rehydration ratio may be performed according to the following calculations.

Rehydration ratio: If weight of the dried sample is 10 g (Wd) and the weight of the sample after rehydration is 60 g (Wr), rehydration ratio is:

$$\frac{Wr}{Wd} = \frac{60}{10} = \frac{6}{1}, \text{or 6 to 1}$$

Rehydration coefficient: The weight of rehydrated sample is 60 g (Wr); the weight of dried sample is 10 g (Wd) and its moisture is 5% (Wu); raw material before drying had 87% water (A); rehydration coefficient is:

$$\frac{Wr}{\frac{Wd - Wu \times 100}{100 - A}} = \frac{60 \times (100 - 87)}{10 - [10 \times 0.05]} = \frac{780}{9.5} = 82.1$$

A simpler test for eating quality can be carried out without weighing and measuring. The material is placed in a cooking pot with water (and a little salt). The pot is then covered and boiled as described above. Except for a few products which are eaten in the dry state, most dried fruit and all dried vegetables are prepared by soaking and cooking. Often this preparation is carried out incorrectly and dried products get a bad reputation.

Good quality dried products, after cooking and if properly treated should be similar to cooked fresh produce. In order to get good results, the following methods are recommended:

- *Quick method*: Cold water, ten times the weight of the dry product, is added to the dried product. The container is covered, brought to the boil and simmered GENTLY until the product is tender. The cooking time may be 15 to 45 minutes after the boiling point has been reached.
- *Slow method*: This gives better results than the quick method. Cold water is added to the dry food and is left to soak for 1 to 2 hours before cooking. The product is then cooked in the same water as that in which it was soaked. The actual cooking time will probably be shorter than that for the quick method.

Other points to remember are:

- if too much water is added the cooked product will have little flavour. However, if too little water is added the product may dry and burn. This can be avoided by adding small quantities of water during cooking;
- always cook with a lid on the container;
- salt, if required, should be added when the cooking is almost complete;
- partly used packages of dry products should be reclosed tightly or kept in containers with good fitting lids.

Handling, Sorting, Packing and Storage of Fruits

It is not easy to assess when drying has been completed. In

absence of instrumentation, the characteristics of the various products after drying / dehydration can only be assessed by experience. Although this cannot be conveyed adequately on paper, some general indications can be given.

Fruit products: When a handful of fruit is squeezed tightly together in the hand and then released, the individual pieces should drop apart readily and no moisture be left behind on the hand. It should not be possible to separate the skin by rubbing unpeeled fruit and the fruit centre should no longer reveal any moist area. Banana should be leathery and not too tough to eat in their dry state.

Vegetable products: Onions should be dried until they are crisp whereas tomatoes should be leathery. In general, the lower the moisture content, the better the keeping quality will be, but overdried products generally have an inferior quality. Also the loss in weight from excessive drying cannot be tolerated in a commercial operation designed to run profitably. It is, however, essential to dry up to an optimum / safe moisture level, related to the type of the product and his designed shelf life, and to avoid running the risk of the products becoming spoiled due to excess water content.

When drying is completed, the material should be sorted either on trays or on a table in order to remove pieces of poor quality and colour and any foreign matter. Very fine material should be separated from the bulk of the material by using a sieve (12 or 16 mesh per inch). Bad quality products which show poor colour need to be removed from the bulk of finished product. After selection and grading, dried products should be packed immediately, preferably in polythene bags which must be folded and closed / tied tightly. However, plastic bags are easily damaged and therefore they must be packed into cartons or jute sacks before they are transported.

Deterioration of Dried Fruit

Dried fruit must be considered as a relatively perishable commodity in the same category as cereals, pulses and similar stored products. It is subject to deterioration resulting from mould growth, insect and mite infestation and physical and chemical changes.

Mould Growth

When the moisture content of dried fruit is allowed to exceed the maximum permissible level for safe storage then mould growth may occur. At the present time, suitable field moisture meters for use with dried fruit are not readily available, and moisture determinations can only be satisfactorily carried out where laboratory facilities are available. Various species of drought resisting fungi may develop on dried fruit when the moisture content is just above the safe level, and a number of osmophilic yeasts are quite commonly associated with spoilage in dried fruit.

Many of the yeasts bring about fermentation with the production of lactic acid or alcohol, and yeasts are frequently present in wart-like crystalline growths which occur in fruit which has become "sugared". In very moist fruit mucoraceous fungi may predominate and are visible as white fluffy growths on and within the fruit.

Mite Infestation

Severe mite infestations are often associated with the growth of osmophilic yeasts in fermenting dried fruit products. Many of these mites are unable to complete their development in the absence of yeast. They have been reported as occurring on dried fruit, and particularly figs and prunes in Mediterranean countries. Such infestations are difficult to eradicate and affect consumer acceptance of the contaminated products.

Insect Infestation

Insect infestations may begin in the field before harvest, may continue during bulk storage after drying, and unless measures are taken to prevent it, may occur in the finished packaged product during storage prior to distribution and consumption. Regular treatments of the stack of dried fruit with a suitable insecticide will be necessary as a routine to combat light insect infestations. Pyrethrins synergised with piperonyl butoxide are commonly used as a surface spray or as an aerosol fog for this purpose. Heavy infestations will require that the fruit be fumigated.

Technology of Semi-processed Fruit Products

The semi-processed fruit products are manufactured in order to be delivered to industry processing centres (in the fruit producing country itself or in importing countries) where they will be further manufactured in consumer oriented finished products: jams, jellies, syrups, fruits in syrup, etc.

In the practice of semi-processed fruit products and for the purpose of this document the following categories are defined:

- *fruit "pulps"*: semiprocessed products, not refined, obtained by mechanical treatment (or, less often, by thermal treatment) of fruit followed by their preservation. Either whole fruit, halves or big pieces are used which enables easy identification of the species. "Pulps" can be classified in boiled or non boiled (raw).
- *fruit "purées-marks"*: semiprocessed products obtained by thermal and mechanical treatment or, very rare, raw and then refined, operations by which all nonedible parts (cores, peels, etc.) are removed. "Purées-marks" are classified in boiled (the more usual case) and non boiled (raw).
- *semiprocessed juices*: products obtained by cold pressure or very rare by other treatments (diffusion, extraction, etc.) followed by the preservation.

Preservation of Semiprocessed Fruit Products

Preservation can be achieved by chemical means, by freezing or by pasteurization. The choice of preservation process for each individual case is a function of the semiprocessed product type and the shelf life needed.

Chemical Preservation

In many countries, in practice, this is carried out with sulphur dioxide, sodium benzoate, formic acid and, on a small scale, with sorbic acid and sorbates. Preservation with sulphur dioxide is a widespread process because of its advantages: universal antiseptic action and very economic application. The drawbacks of SO_2 are: SO_2 turn firms the texture of some fruit species (pomaces), desulphiting is not always complete and recolouring of red fruits is not always complete after desulphitation. Practical preservation dosage levels with SO_2 for about 12 months is 0.18-0.20% SO2 (with respect to the product to be preserved). This level could be reduced to 0.09% SO_2 for 3 months and to 0.12% SO_2 for 6 months preservation. The preservation with sulphur dioxide is in use mainly for "pulps" and for "purées-marks".

Chemical preservation can be performed from a practical point of view by the utilisation of 6% SO_2 water solutions or by direct introduction of sulphur dioxide gas in the product (for "purées-marks"). The preparation of 6% SO_2 solutions is done by bubbling the gas from cylinders in cold water; from a 50 kg SO_2 compressed gas cylinder results 830 l of 6% SO_2 solution. These SO_2 solutions have to be stored in cool places, in closed receptacles and with periodic concentration control/check by titration or by density measurements approximate results.

Preservation with sodium benzoate has the following advantages: it does not firm up the texture and does not modify fruit colour. The disadvantages are: it is not a universal antiseptic, its action needs an acid medium and the removal is partial. Sodium benzoate is in use for "pulps" and for "purées-marks" but less for fruit juices. Practical dosage level for 12 months preservation is 0.18-0.20 % sodium benzoate, depending on the product to be preserved. Sodium benzoate is used as a solution in warm water; the dissolution water level has to be at maximum 10% reported to semiprocessed product weight.

Formic acid preservation is performed mainly for semiprocessed fruit juices at a dosage level of 0.2 % pure formic acid (100%). Formic acid is an antiseptic effective against yeasts, does not influence colour of products and is easily removed by boiling. Formic acid could be diluted with water in order to insure a homogeneous distribution in the product to be preserved; water has to be at maximum 5 % of the product weight. Because of a potential effect of pectic substance degradation, formic acid is less in use for "pulps" and "purées-marks" preservation.

Sorbic acid used as potassium sorbate can be used for preservation of fruit semiprocessed products at a dosage level of 0. 1% maximum. Advantages of sorbates are: they are completely harmless and without any influence on the organoleptic properties of semiprocessed fruit products.

Preservation by Pasteurisation

As fruit has a low pH, preservation of semiprocessed fruit products could also be performed by pasteurisation, the length of this step varying with the size of the receptacles. The advantages of this type of treatment are: hygienic process, which assure a long term preservation; the disadvantages are: need for air tight receptacles, and pectic substances could begin to deteriorate if the thermal treatment is too long.

Thermal preservation of fruit semiprocessed products could also be done by a "selfpasteurization": very hot semiprocessed products are filled into receptacles which are sealed and then inverted in order to sterilise the air which goes through the hot fruit mass.

Preservation by Freezing

This is done on an industrial scale in some countries and can be done with or without sugar addition. The advantages of this process are: absence of added substances; very good preservation of quality of fruit constituents (pectic substances, vitamins, etc.) and good preservation of organoleptic properties (flavour, taste, colour). Freezing is done at about -20 to -30° C and storage at -10 to -18° C. Freezing is applied mainly to semiprocessed fruit products aimed at very high quality and high cost finished products.

Chemical Preservation

Sorting is needed in order to remove sub-standard fruit (with moulds, with diseases, etc.) and all foreign bodies.

Washing is obligatory in order to remove all impurities which cannot be eliminated at the processing step in finished products.

Coring and Cutting, mainly for pomace fruits, has as main objective a better utilisation of preservation "space" in receptacles and is not mandatory; this will be defined by customer/ supplier agreements / standards. This operation is preferably performed by mechanical means.

Preservation is carried out with the 6% SO_2 solution which is added to the prepared fruits (placed in bulk in receptacles) in the quantity needed to obtain the preservation dosage level. For a better / homogeneous preservative distribution, the initial 6% SO_2 solution could be diluted with water; however, the diluted solution (which will be filled in receptacles) has to be at a dosage level of less than 10% of the semiprocessed product weight.

For some soft fruit, especially strawberries, preservation is done with a mix of 6% SO_2 solution and calcium bisulphite solution (containing also 6% SO_2). Preparation of calcium bisulphite solution is done by the introduction of 30 kg of CaO in 1 m^3 SO_2 solution and mixing up to clarification. The resulting solution is mixed with the initial 6% SO_2 solution, generally in a 1:1 ratio, but the ratio can be adapted to the fresh fruit texture. Firming of soft fruit texture by this treatment is based on the formation of calcium pectate with pectic substances from fruit tissues.

In the case of sodium benzoate, formic acid or potassium sorbate, the dosage levels to be used are as indicated above with the rule that it is not allowed to add more that 10% liquid in receptacles on the prepared fruits. Preservation by pasteurisation or "self-pasteurisation" will need as additional steps: a) boiling with a minimum water addition (maximum 10%); b) filling of receptacles; c) hermetic closing followed by d) pasteurisation or "self-pasteurisation".

Fruit Sugar Preserves Technology

As a overall rule of thumb, a sugar concentration of about 60% in finished or processed fruit products generally insures their preservation. Preservation is not only determined by the osmotic pressure of sugar solutions but also by the water activity values in the liquid phase, which can be lowered by sugar addition; and by evaporation down to 0.848 aw; this value however does not protect products from mould and osmophile yeast attack. Maximum saccharose concentration that can be achieved in the liquid phase of the product is 67.89%; however higher total sugar quantities (up to 70-72%) found in products are explained by an increased reducing sugar solubility resulting from saccharose inversion.

Jams

The preservation of fruit by jam making is a familiar process carried out on a small scale by housewives in many parts of the world. Factory jam making has become a highly complex operation, where strict quality control procedures are employed to ensure a uniform product, but the manufacturing operations employed are in essence the same as those employed in the house.

Fresh or pre-cooked fruit is boiled with a solution of cane or beet sugar until sufficient water has been evaporated to give a mixture which will set to a gel on cooling and which contains 32-34% water. Gel formation is dependent on the presence in the fruit of the carbohydrate pectin, which at a pH of 3.2—3.4 and in the presence of a high concentration of sugar, has the property of forming a viscous semi-solid.

During jam boiling, all microorganisms are destroyed within the product, and if it is filled hot into clean receptacles which are subsequently sealed, and then inverted so that the hot jam contacts the lid surface, spoilage by microorganisms will not take place during storage. About 30% of the vitamin C present in fresh fruit is destroyed during the jam-making process, but that which remains in the finished product is stable during storage.

The high moisture content of jam makes it susceptible to mould damage once the receptacle has been opened and exposed from some time to the air. No problems of microbiological spoilage are likely to arise in the canned product during storage.

Marmalade

This sugar preserve is defined as "semisolid or gel-like product prepared from fruit ingredients together with one or more sweetening ingredients and may contains suitable food acids and food pectins; the ingredients are concentrated by cooking to such a point that the TSS—Total Soluble Solids—of the finished marmalade is not below 65%".

Fruit Paste

Fruit paste is a product obtained in the same way as special non-gelified fruit marmalade but with a lower water content—about 25% TSS in fruit paste. Lowering water content could be achieved by continuing boiling of the product or by drying the product by natural or artificial drying. An example of paste without sugar is the sun dried apricot or prune paste.

Preparation of Jams, Jellies and Marmalade

- Boil the pulp or the juice (with water when necessary)
- Add the pectin
  - to the batch while stirring very vigorously
  - Pectin which has previously been mixed with 5 times its weight in sugar taken from the recipe)
- Boil for about 2 minutes to assure a complete dissolution
- Add the sugar while keeping the batch boiling
- Boil down quickly to desired Brix
- Add the acid (usually citric acid) and remove the froth
- Fill hot into the (previously cleaned) jars and close
- Invert the jars for three minutes to pasteurise the cover

Pineapple-papaya Jam

The fruit should be prepared as per previous instructions. For pineapples, the ends are removed and discarded; the cores and outer parts of the fruits are also removed. The fruit cylinders obtained are pulped through a special extractor (Fitzpatrick communiting machine) equipped with a 0.40-in screen sieve; the pulp thus obtained is used for making jam. The papaya are prepared by hand-peeling the fruit; the fruit is then halved and the seeds removed. It is then pulped in the communiting machine using a 0.40-in screen sieve. When ginger root is used as flavouring, it is peeled and macerated in a Kenwood blender to a very fine consistency.

A typical formula for a pineapple-papaya jam (50:50 ratio) with ginger flavouring is given as follows:

Pineapple pulp	25.0
Papaya pulp	25.0 pounds
Cane sugar	50.0
Apple pectin (150 grade)	6.0 ounces
Citric acid	6.4
Fresh ground ginger	7.5

Processing

The weighed fruit pulp is placed in a stainless steel steam-jacketed kettle and heated to about 110°F under constant stirring. When the product reaches this temperature, the heat is turned off. The pectin (mixed in about ten times its weight with some of the weighed sugar), is then mixed into the fruit pulp, stirring constantly in order to prevent the pectin from clotting.

When the pectin has dissolved, the remainder of the sugar is added and dissolved completely in the mixture. The heat is then turned on and the jam mixture is stirred constantly until it starts boiling vigorously. During the remainder of the cooking, the product is stirred occasionally. Near the finishing point (approximately 221° F), the citric acid and the ginger (if it is used) are also added.

Determination of the finishing point is done by removing samples at intervals, cooling, and reading the soluble solids by means of a refractometer equipped with a Brix scale. After the jam reaches the proper Total Soluble Solids content, the heat is turned off and the surface scum/foam is removed.

The jam then is quickly put into receptacles which have been cleaned and sterilised with boiling in water for 30 minutes. The filling operation is done rapidly in order to prevent the temperature of the jam from falling below 190° F. After filling, sterilised lids (boiled for 30 minutes in water) are placed on the receptacles and they are then sealed. After this operation the receptacles are inverted for about 3 minutes to insure that the lids are sterilised. The receptacles are then placed upright. At this stage it is not necessary to do any further processing, therefore the receptacles are cooled in running cold water until they reach a temperature slightly above room temperature. They are then dried in air and labelled.

Finished Products Evaluation

During production at medium / large scale, it is recommended that quality controls be performed during manufacturing. After ten weeks of storage at room temperature it is recommended that an examination of finished products be performed. The receptacles are opened and contents carefully emptied on to enamel trays without disturbing the formation of the jam. The empty cans are then inspected for signs of corrosion. Factors other than flavour include colour, appearance, syrup separation, firmness and spreading quality. For flavour, jam is tested on pieces of bread.

Samples are taken for measurement of pH (with a glass electrode pH meter) and Total Soluble Solids (with a refractometer equipped with a Brix scale). This evaluation enables to have a quality check during product shelf life and to obtain data needed for necessary improvements of future productions. For pineapple-papaya jam, products made with 30% pineapple and 70% papaya with added ginger has the highest score for flavour. The use of plain tin cans causes corrosion problems which is not the case when acid resistant lacquer cans are used

Technology of Fruit Juices

Fruit juices are products for direct consumption and are obtained by the extraction of cellular juice from fruit, this operation can be done by pressing or by diffusion. For the purpose of this document, the technology of fruit juice processing will cover two finished product categories:

— juices without pulp ("clarified" or "not clarified");

— juices with pulp ("nectars").

We will also define as:

— "natural juices" products obtained from one fruit; and

— "mixed juices" products obtained from the mix of two or three juices from different fruit species or by adding sugar.

Juices obtained by removal of a major part of their water content by vacuum evaporation or fractional freezing will be defined as "concentrated juices".

Technological Steps for Processing of Fruit Juices without Pulp

Fruit juices must be prepared from sound, mature fruit only. Soft fruit varieties such as grapes, tomatoes and peaches should only be transported in clean boxes which are free from mould and bits of rotten fruit.

Washing: fruit must be thoroughly washed. Generally, fruit will be submitted to a pre-washing before sorting and a washing step just after sorting.

Sorting: removal of partially or completely decayed fruit is the most important operation in the preparation of fruit for production of first quality fruit juices; sorting is carried out on moving inspection belts or sorting tables.

Crushing/Grinding/Disintegration Step is applied in different ways and depends on fruit types:

— Crushing for grapes and berries;

— Grinding for apples, pears;

— Disintegration for tomatoes, peaches, mangoes, apricots.

This processing step will need specific equipment which differs from one type of operation to another.

Enzyme Treatment of crushed fruit mass is applied to some fruits by adding 2-8% pectolitic enzymes at about 50° C for 30 minutes. This optional step has the following advantages: extraction yield will be improved, the juice colour is better fixed and finished product taste is improved. However, for fruit which is naturally rich in pectic substances, this treatment makes the resulting "exhausted" material useless for industrial pectin production.

Heating of crushed fruit mass before juice extraction is an optional step used for some fruit in order to facilitate pressing and colour fixing; at same time, protein coagulation takes place.

Pressing to extract juice.

Diffusion is an alternative step for juice extraction and can be carried out discontinuously or in batteries at water temperature of about 80-85 ° C.

Juice Clarifying can be performed by centrifugation or by enzyme treatment. Centrifugation achieves a separation of particles in suspension in the juice and can be considered as a pre-clarifying step. This operation is carried out in centrifugal separators with a speed of 6000 to 6500 RPM. Enzyme clarifying is based on pectic substance hydrolysis; this will decrease the juices' viscosity and facilitate their filtration.

The treatment is the addition of pectolitic enzyme preparations in a quantity of 0.5 to 2 g/l and will last 2 to 6 hours at room temperature, or less than 2 hours at 50° C, a temperature that must not be exceeded. The control of this operation is done by checking the decrease in juice viscosity. Sometimes, the enzyme clarifying is completed with the step called "sticking" by the addition of 5-8 g/hl of food grade gelatine which generates a flocculation of particles in suspension by the action of tannins.

Filtration of clarified juice can be carried out with kieselgur and bentonite as filtration additive in press-filters (equipment).

De-Tartarisation is applied only to raisin juice and is aimed to eliminate potassium bitartrate from solution. This step can be performed by the addition of 1% calcium lactate or 0.4% calcium carbonate. Pasteurisation of juice can be done for temporary preservation (pre-pasteurisation) and in this case this operation is carried out with continuous equipment (heat exchangers, etc.); warm juice is stored in drums or large size receptacles (20-30 kg). Pasteurisation conditions are at 75°C in continuous stream.

Pasteurisation of bottled juice is then carried out just before delivery to the market; this is performed in water baths at 75° C until the point where the juice reaches 68° C. In cases when the final pasteurisation is done without pre-pasteurisation and temporary storage, modern methods use a rapid pasteurisation followed by aseptic filling in receptacles. Rapid pasteurisation conditions are as follows: temperature about 80° C, over 10-60 sec., followed by cooling; all operations are carried out in continuous stream.

Preservation under CO_2 pressure may be done at a concentration of 1.5% CO_2 under a pressure of 7 kg/cm^2. At the distribution step, proceed at CO_2 decompression and the juice is then submitted to a sterilising filtration and aseptic filling in receptacles. Preservation by freezing is carried out at about -30° C, after a preliminary de-aeration; storage is at -15 to -20° C. Production of concentrated juices by evaporation is performed under vacuum (less than 100 mm Hg residual pressure) up to a concentration of 65-70% total sugar which assures preservation without further pasteurisation.

Modern evaporation installations recover flavours from juices which are then reincorporated in concentrated juices. Additional operations for juice manufacturing are the vacuum de-aeration and mixing with other fruit juices or with sugar. For the production of non clarified juices the centrifugation is the only specific step, enzyme clarifying and subsequent filtration being eliminated. The optimum sugar/acid ratio for the majority of fruit, mainly for pomaces, is 10/1 to 15/1. Fruit which is rich in carotenoids is only processed as juices with pulp ("nectars").

Technological Flow-sheet for Nectars

This process is divided at industrial scale in two categories of operations:

— processing for obtaining juices;

— juice conditioning for preservation.

Operations in the first category are differ according to the type of fruit which to be processed: Pomaces (apples, pears) are washed and sorted and then crushed in a colloid mill; fruit purée is then passed through a screw type heating equipment where direct steam is used as a source of heat. Warm fruit mass is treated in a pulper with a 2 mm screen and then through an extractor similar with the equipment used for tomato juice.

Stone fruits (apricots, peaches, cherries, etc.) after washing and sorting are submitted to steam in a continuous heater, then the warm fruit mass is passed through a pulper and then an extractor (as mentioned above). Berries (strawberry, wild berries, etc.) are washed, sorted and then crushed, preheated and then introduced in extractor. In order to avoid browning and undesirable taste modifications it is usual to add about 0.05% ascorbic acid.

Second category type of operations are similar for all fruit species: Partial elimination of cellulose is achieved with a continuous centrifugal separator; the resulting juice is then processed in order to adjust sugar and acid content for viscosity. Sugar (about 8-10%) is added as a syrup (in water or in the juice of same fruit obtained by pressure). Acidity is adjusted with citric or tartaric acid. The adjusted juice is then deaerated under vacuum at about 40° C. This step aims at avoiding oxidative reactions and vitamin C loss reduction.

An important subsequent step is an intensive homogenisation (under pressure at 150-180 A) in order to obtain particles with dimensions below 100. The homogenised juice obtained is then continuously pasteurised in plate heat exchanger equipment at a temperature of about 130° C, cooled down to about 90° C and aseptically packed in receptacles. The principal characteristics of fruit "nectars" are uniformity and stability of the content provided by the advanced disintegration of fruits.

Stability can be obtained by increasing product viscosity by adding pectin for fruit which is deficient in this component. In order to avoid "separation", intensive homogenisation is carried out as described above. Fruit "nectars" contain all the important components of the original fruit and to a large extent maintain their taste and flavour. The sugar/acidity (as citric acid) ratio is to a large extent determined by the type of fruit and the correction applied; for example, this ratio is 30 for apricots, 40 for peaches, 160 for pears, etc.

References

Beychok, M.R., *Fundamentals Of Stack Gas Dispersion*, 4th Edition, author-published, 2005.

Brennan, J. G., Butters, J. R., Cowells, N. D. and Lilly, A. E., *Food Engineering Operations*, 3rd Edition, Applied Science, London, 1990.

Charm, S. E., *The Fundamentals of Food Engineering*, 2nd Edition, AVI, Westport, 1971.

Coulson, J. M. and Richardson, J. F., *Chemical Engineering*, 3rd Edition, Vols. 1, 2, Pergamon, Oxford, 1977, 1978.

Danny D. Reible., *Fundamentals of Environmental Engineering*, CRC Publishers, 1998.

Emken, E. A., "Nutrition and biochemistry of trans and positional fatty acid isomers in hydrogenated oils," *Annual Reviews of Nutrition*, 4: 339-376, 1984.

Fellows, P., *Food Processing Technology, Principles and Practice*, 2nd Edition, Woodhead, Cambridge, 2000.

James R. Mihelcic, Martin T. Auer, and others., *Fundamentals of Environmental Engineering*, John Wiley, 1999.

Loncin, M. and Merson, R. L. *Food Engineering; Principles and Selected Applications*, Academic, New York, 1979.

Tchobanoglous, G., Burton, F.L., and Stensel, H.D., *Wastewater Engineering (Treatment Disposal Reuse) / Metcalf & Eddy, Inc.*, 4th Edition, McGraw-Hill Book Company, 2003.

Turner, D.B., *Workbook of atmospheric dispersion estimates: an introduction to dispersion modeling*, 2nd Edition, CRC Press, 1994.

5

Vegetable Specific Processing Technologies

Vegetable processors must appreciate the substantial differences that varieties of a given vegetable will possess. In addition to variety and genetic strain differences with respect to weather, insect and disease resistance, varieties of a given vegetable will differ in size, shape, time of maturity, and resistance to physical damage. Varietal differences then further extend into warehouse storage stability, and suitability for such processing methods as canning, freezing, pickling or drying.

A variety of peas that is suitable for canning may be quite unsatisfactory for freezing and varieties of potatoes that are preferred for freezing may be less satisfactory for drying or potato chip manufacture. This should be expected since different varieties of a given vegetable will vary somewhat in chemical composition, cellular structure and biological activity of their enzyme system.

Harvesting and Pre-processing

When vegetables are maturing in the field they are changing from day to day. There is a time when the vegetable will be at peak quality from the stand-point of colour, texture and flavour. This peak quality is quick in passing and may last only a day. Harvesting and processing of several vegetables, including tomatoes, corn and peas are rigidly scheduled to capture this peak quality.

After the vegetable is harvested it may quickly pass beyond the peak quality condition. This is independent of microbiological spoilage; these main deteriorations are related to:

— loss of sugars due to their consumption during respiration or their conversion to starch; losses are slower under refrigeration but there is still a great change in vegetable sweetness and freshness of flavour within 2 or 3 days;

— production of heat when large stockpiles of vegetables are transported or held prior to processing. At room temperature some vegetables will liberate heat at a rate of 127,000 kJ/ton/day; this is enough for each ton of vegetables to melt 363 kg of ice per day. Since the heat further deteriorates the vegetables and speeds microorganisms growth, the harvested vegetables must be cooled if not processed immediately. But cooling only slows down the rate of deterioration, it does not prevent it, and vegetables differ in their resistance to cold storage. Each vegetable has its optimum cold storage temperature which may be between about 0-100 C (32-500 F).

— the continual loss of water by harvested vegetables due to transpiration, respiration and physical drying of cut surfaces results in wilting of leafy vegetables, loss of plumpness of fleshy vegetables and loss of weight of both.

Moisture loss cannot be completely and effectively prevented by hermetic packaging. This was tried with plastic bags for fresh vegetables in supermarkets but the bags became moisture fogged, and deterioration of certain vegetables was accelerated because of buildup of CO_2 and decrease of oxygen in the package. It therefore is common to perforate such bags to prevent these defects as well as to minimise high humidity in the package which would encourage microbial growth.

Shippers of fresh vegetables and vegetable processors, whether they can, freeze, dehydrate, or manufacture soups or ketchup, appreciate the instability and perishability of vegetables and so do everything they can to minimise delays in processing of the fresh product. In many processing plants it is common practice to process vegetables immediately as they are received from the field.

To ensure a steady supply of top quality produce during the harvesting period the large food processors will employ trained field men; they will advise on growing practices and on spacing of plantings so that vegetables will mature and can be harvested in rhythm with the processing plant capabilities. This minimises stockpiling and need for storage.

Cooling of vegetables in the field is common practice in some areas. Liquid nitrogen-cooled trucks may next provide transportation of fresh produce to the processing plant or directly to market. Upon arrival of vegetables at the processing centre the usual operations of cleaning, grading, peeling, cutting and the like are performed using a moderate amount of equipment but a good deal of hand labour also still remains.

Reception

This covers qualitative and quantitative control of delivered vegetables. The organoleptic control and the evaluation of the sanitary state, even if they are very important steps in vegetables' characteristics assessment, cannot establish their technological value. On the other hand, laboratory controls do not precisely establish their technological properties because of the difficulty in putting into showing some deterioration when using rapid control methods.

One correct method of vegetable quality appraisal is their overall evaluation based on the whole complex of data that can be obtained by combining an extensive organoleptic evaluation with simple analysis that can be performed rapidly in plant laboratory. These analysis can be:

— refractometric extract (tomatoes, fruit, etc.);

— specific weight (potatoes, peas, etc.);

— consistency (measured with tenderometers, penetrometers, etc.);
— boiling tests, etc.

Temporary Storage

This step should be as short as possible and better completely eliminated. Vegetables can be stored in:

— simple stores, without artificial cooling;
— in refrigerated stores; or, in some cases,
— in silos (potatoes, etc.).

Simple stores should be covered, fairly cool, dry and well ventilated but without forced air circulation which can induce significant losses in weight through intensive water evaporation; air relative humidity should be at about 70-80%. Refrigerated storage is always preferable and in all cases a processing centre needs a cold room for this purpose.

Washing Equipment

Washing is used not only to remove field soil and surface microorganisms but also to remove fungicides, insecticides and other pesticides, since there are laws specifying maximum levels of these materials that may be retained on the vegetable; and in most cases the allowable residual level is virtually zero. Washing water contains detergents or other sanitisers that can essentially completely remove these residues.

The washing equipment, like all equipment subsequently used, will depend upon the size, shape and fragility of the particular kind of vegetable:

— flotation cleaner for peas and other small vegetables;
— rotary washer in which vegetables are tumbled while they are sprayed with jets of water; this type of washer should not be used to clean fragile vegetables;

Sorting

This step covers two separate operations:

— removal of non-standard vegetables (and fruit) and possible foreign bodies remaining after washing;
— quality grading based on variety, dimensional, organoleptical and maturity stage criterion.

Skin Removal

Some vegetables require skin removal. This can be done in various ways.

Mechanical

This type of operation is performed with various types of equipment which depend upon the result expected and the characteristics of the fruit and vegetables, for example:

— a machine with abrasion device (potatoes, root vegetables);
— equipment with knives (apples, pears, potatoes, etc.);
— equipment with rotating sieve drums (root vegetables). Sometimes this operation is simultaneous with washing (potatoes, carrots, etc.) or preceded by blanching (carrots).

Chemical

Skins can be softened from the underlying tissues by submerging vegetables in hot alkali solution. Lye may be used at a concentration of about 0.5-3%, at about 93° C (2000 F) for a short time period (0.5-3 min). The vegetables with loosened skins are then conveyed under high velocity jets of water which wash away the skins and residual lye. In order to avoid enzymatic browning, this chemical peeling is followed by a short boiling in water or an immersion in diluted citric acid solutions. It is more difficult to peel potatoes with this method because it is necessary to dissolve the cutin and this requires more concentrated lye solutions, up to 10%.

Thermal

Wet heat (steam): Other vegetables with thick skins such as beets, potatoes, carrots and sweet potatoes may be peeled with steam under pressure (about 10 at) as they pass through cylindrical vessels. This softens the skin and the underlying tissue. When the pressure is suddenly released, steam under the skin expands and causes the skin to puff and crack. The skins are then washed away with jets of water at high pressure (up to 12 at).

Dry heat (flame): Other vegetables such as onions and peppers are best skinned by exposing them to direct flame (about 1 min at 1000° C) or to hot gases in rotary tube flame peelers. Here too, heat causes steam to develop under skins and puff them so that they can be washed away with water.

Blanching Treatment

The special heat treatment to inactivate enzymes is known as blanching. Blanching is not indiscriminate heating. Too little is ineffective, and too much damages the vegetables by excessive cooking, especially where the fresh character of the vegetable is subsequently to be preserved by processing. This heat treatment is applied according to and depends upon the specificity of vegetables, the objectives that are followed and the subsequent processing / preservation methods.

Two of the more heat resistant enzymes important in vegetables are catalase and peroxidase. If these are destroyed then the other significant enzymes in vegetables also will have been inactivated. The heat treatment to destroy catalase and peroxidase in different vegetables are known, and sensitive chemical tests have been developed to detect the amounts of these enzymes that might survive a blanching treatment.

Because various types of vegetables differ in size, shape, heat conductivity, and the natural levels of their enzymes, blanching treatments have to be established on an experimental basis. As with sterilisation of foods in cans, the larger the food item the longer it takes for heat to reach the centre. Small vegetables may be adequately blanched in boiling water in a minute or two, large vegetables may require several minutes. Blanching as a unit operation is a short time heating in water at temperatures of 100° C or below. Water blanching may be performed in double bottom kettles, in special baths with conveyor belts or in modern continuous blanching equipment.

In order to reduce losses of hydrosoluble substances (mineral salts, vitamins, sugars, etc.) occurring during water blanching, several methods have been developed:

- temperature setting at 85-95° C instead of 100° C;
- blanching time has to be just sufficient to inactivate enzymes catalase and peroxidase;
- assure elimination of air from tissues.

Steam heat treatment can also be applied instead of water blanching as a preliminary step before freezing or drying, as long as the preservation method is only used for enzyme inactivation and not to modify consistency. For drying, the vegetables are conveyed directly from steaming equipment to drying installations without cooling. Vegetable steaming is carried out in continuous installations with conveyer belts made from metallic sieves.

Cooling of vegetables after water blanching or steaming is performed in order to avoid excessive softening of the tissues and has to follow immediately after these operations; one exception is the case of vegetables for drying which can be transferred directly to drying equipment without cooling. Natural cooling is not recommended because is too long and generates significant losses in vitamin C content. Cooling in pre-cooled air is sometimes used for vegetables that will be frozen

Cooling in water can be achieved by sprays or by immersion; in any case the vegetables have to reach a temperature value under 37° C as soon as possible. Too long a cooling time generates supplementary losses in valuable hydrosoluble substances; in order to avoid this, the temperature of the cooling water has to be as low as possible.

Canning Operation

Large quantities of vegetable products are canned. A typical flow sheet for a vegetable canning operation (which also applies to fruit for the most part) covers some food process unit operations performed in sequence: harvesting; receiving; washing; grading; heat blanching; peeling and coring; can filling; removal of air under vacuum; sealing/closing, retorting/heat treatment; cooling; labelling and packing. The vegetable may be canned whole, diced, puréed, as juice and so on.

On-line Simplified Methods

Peroxidase Test

- *Solutions*: In order to check the peroxidase activity two solutions have to be prepared:
  - 1% guaiacol in alcohol solution (1 g guaiacol is dissolved in about 50 cm^3 of 96% ethylic alcohol and then this preparation is brought to 100 cm° with the same solvent);
  - peroxide solution 0.3% (1 cm^3 perhydrol is brought to 100 cm^3 with distilled water.
- *Sampling*: From various parts of the material samples are taken (about 20-30 pieces, etc.); the material is then crushed in a laboratory bowl in order to obtain an average sample.
- *Check*: Prom the average sample, 10-20 g of material is taken in a medium capacity test tube; on this sample are poured: 20 cm^3 distilled water; 1 cm^3 of 1% guaiacol solution; 1.6 cm^3 of peroxide solution.

The contents of the test tube is shaken well. The gradual appearance of a weak pink colour indicates an incomplete peroxidase inactivation—reaction slightly positive. If there are no tissue colour modifications after 5 minutes, the reaction is negative and the enzymes have been inactivated. As an orientative check it is also possible to simply pour a few drops of 1% guaiacol solution and 0.3% peroxide solution directly on blanched and crushed vegetables. A rapid and intensive brown-reddish tissue colouring indicates a high peroxidase activity (positive reaction).

Catalase Test

In order to identify the catalase enzyme activity, 2 g of dehydrated vegetables are well crushed and mixed with about 20 cm^3 of distilled water. After 15 min softening, 0.5 cm^3 of a 0.5% or 1% peroxide solution is poured on prepared vegetables. In the presence of catalase, a strong oxygen generation is observed for about 2-3 minutes. These tests are of a paramount importance in order to determine the vegetable blanching treatments (temperature and time); incomplete enzyme inactivation has a negative effect on finished product quality. For cabbage catalase inactivation by blanching is sufficient; blanching further to peroxidase inactivation would have negative effects on product quality and even complete browning. For all other vegetables and for potatoes, both tests must be negative, for catalase and for peroxidase.

Fresh Vegetable Storage

The vegetables can be stored, in some specific natural conditions, in fresh state, that is without significant modifications of their initial organoleptic properties. Fresh vegetable storage can be short term; this was briefly covered under temporary storage before processing. Also fresh vegetable storage can be long term during the cold season in some countries and in this case it is an important method for vegetable preservation in the natural state.

In order to assure preservation in long term storage, it is necessary to reduce respiration and transpiration intensity to a minimum possible; this can be achieved by:

— maintenance of as low a temperature as possible (down to 0° C),

— air relative humidity increased up to 85-95 % and

— CO_2 percentage in air related to the vegetable species.

Vegetables for storage must conform to following conditions: they must be of one of the autumn or winter type variety; be at edible maturity without going past this stage; be harvested during dry days; be protected from rain, sun heat or wind; be in a sound state and clean from soil; be undamaged. From the time of harvest and during all the period of their storage vegetables are subject to respiration and transpiration and this is on account of their reserve substances and water content.

The more the intensity of these two natural processes are reduced, the longer sound storage time will be and the more losses will be reduced. For this reason, vegetables have to be handled and transported as soon as possible in the storage conditions (optimal temperature and air relative humidity for the given species). Even in these optimal conditions storage will generate losses in weight which are variable and depend upon the species.

Vegetable Drying

Vegetable Powder Processing Technology

This technology has been developed in recent years with applications mainly for potatoes (flour, flakes, granulated), carrots (powder) and red tomatoes (powder). In order to obtain these finished products there are two processes:

— drying of vegetables down to a final water content below 4% followed by grinding, sieving and packing of products;

— vegetables are transformed by boiling and sieving into purées which are then dried on heated surfaces (under vacuum preferably) or by spraying in hot air.

Industrial installations that can be used for these products and technological data are summarised below:

— Dryers with plates under vacuum are equipped with plates heated with hot water. Stainless steel plates containing the purée to be dried are placed on them. Process conditions are at low residual pressure (about 10-20 mm Hg) and a product temperature of 50-70° C. This equipment is discontinuous but easy to operate.

— Drum dryers have one or two drums heated with hot water or steam as heating elements. Feeding is continuous between the two drums which are rotating in reverse direction (about 2-6 rotations per minute) and the distance of which is adjustable and determines the thickness of layer to be dried the product is dried and removed by mechanical means during rotation.

— Drying installations by spraying in hot air; the product is introduced in equipment and sprayed by a special device in hot air. Drying is instantaneous (1/50 s) and therefore can be carried out at 130-15O° C.

Packing and Storage of Dried Vegetables

Dried vegetables can suffer significant modifications that bring about their deterioration during storage. The factors that determine these degradations impose at same time the type of packaging materials and storage conditions for packaged products. The moisture content of dried vegetables is not constant because of their hygroscopicity and is always in equilibrium with relative humidity of air in storage rooms. Technical solutions for maintaining a low dehydrated products moisture are:

— storage in stores with air relative humidity below 78%;

— use packages that are water vapour proof. The most efficient packages are tin boxes or drums (mainly for long term storage periods); combined packages (boxes, bags, etc.) from complexes (carton with metallic sheets, plastic materials, etc.) mainly for small packages. One solution for some dried vegetables may be the use of waterproof plywood drums.

Modern solutions are oriented not only to the maintaining product moisture during storage but also reducing this parameter by the use of desiccants (substances which absorb moisture) introduced in packages, hermetically closed. A desiccant in current use is calcium oxide. Granulated calcium oxide is introduced in small bags from a material which is permeable to water vapour but which does not permit the desiccant to escape into products. With desiccants, product moisture can be reduced to even below 4%, and this inhibits or reduces the biochemical and microbiological processes during storage.

Another factor that can deteriorate dried/dehydrated vegetables is atmospheric oxygen through the oxidative phenomena that it produces. In order to eliminate the action of this agent some packing methods under vacuum or in inert gases are in use, applied mainly for packing dried carrots in order to avoid beta-carotene oxidation in beta-ionone. In order to avoid the action of oxygen it is also possible to add ascorbic acid as antioxidant.

Sun or artificial light action on dehydrated vegetables generally causes discoloration which can be avoided by using opaque packaging materials. Dehydrated vegetable compression (especially for roots) to form blocks with a weight of 50-600 g, is practiced sometimes; it has as advantages the reduction of evaporation surface and contact with atmospheric oxygen and volume reduction.

Dehydrated vegetables are compressed at about 300 at. Compressed blocks are packaged in heat sealed plastic materials. Storage temperature has an important role because this reduces or inhibits the speed of all physico-chemical, biochemical and microbiological processes, and thus prolongs storage period. The storage temperature should be below 25° C; lower temperatures help maintain taste, colour and water rehydration ratio and also, to some extent, vitamin C.

Potato Crisp Processing

The most important steps involved in potato crisps processing are:

— Selecting, procuring and receiving potatoes
— Storage of potato stock under optimum conditions
— Peeling and trimming the tubers
— Slicing
— Frying in oil
— Salting or applying flavoured powders
— Packaging

Table 1. Moisture and shipping factors for some dehydrated vegetables

Product	Form/cut	Moisture %	Weight kg/m³
Bean (green)	20 nun cut	5	1.6
Bean (lima)	5	3.3	
Beet	6 mm strips	5	1.6-1.9
Cabbage	6-12 mm shreds	4	0.7-0.9
Carrots	5-8 mm strips	5	3-5
Celery	Cut	4	
Garlic	Cloves	4	
Okra	6 mm slices	8	
Onion	Slices	4	0.4- 0.6
Pea (fresh)	Whole	5	3.4
Pepper (hot)	Ground	5	
Pepper (sweet)	5 mm strips	7	
Potato (Irish)	5-8 mm strips	6	2.9-3.2
	Diced	5	3.3-3.6
Tomato	7-10 mm slices	35	

Selection and Storage

It is important to select potatoes of high specific gravity since this characteristic indicates superior yield and lower oil absorption. It is even more important to select potatoes with low reducing sugar contents or to store them at temperatures conducive to the minimising of these substances. Sprouting and fungal damage must also be minimised by the storage conditions.

Peeling Operation

The ideal peeling operation should only remove a very thin outer layer of the potato, leaving no eyes, blemishes, or other material for later removal by hand trimming. It should not significantly change the physical or chemical characteristics of the remaining tissue. Preferably peeling should use small amount of water and result in minimal effluent; compromises will have to be made in all of these aspects of peeling. First, the potatoes are thoroughly washed, not only for sanitary reasons, but also to prevent dirt of grit from abrading the equipment the tubers will dater contact.

Washing may take place in streams, as the potatoes are being conveyed by water streams, or in equipment provided with means for scrubbing the potato with brushes or rubber rolls. In barrel-type washers, potatoes are cleaned by being tumbled and rubbed against each other and against the sides of the barrel while they are immersed in, or sprayed with, water. After washing, the potatoes are allowed to drain, usually on mesh conveyors, and they travel over an inspection belt where foreign material and defective tubers are removed.

The more common peeling methods are abrasion, lye immersion, and steam. Abrasion peelers which may be either batch or continuous, use disks or rollers coated with grit to grind away the potato surface. An important design feature is to ensure that all surfaces of the tuber are equally exposed to the rasping action.

The peel fragments are flushed out of the unit by water sprays. Such systems work best with uniform, round, undamaged potatoes. Some of the advantages of abrasion peelers are their simplicity, compactness, low cost, and convenience. They are particularly suitable for peeling potatoes intended for chipping, since they do not chemically alter the surface layers. About 10% of the original tuber weight is lost through abrasion peeling prior to chipping.

Slicing

The peeled potatoes are cut into slices from 1/15 to 1/25 in. by rotary slicers. Centrifugal force presses the tuber against stationary gauging shoes and knives. Thickness is varied, not only to meet consumer preferences, but also to fit the condition of the tubers and the frying temperature and time. Slices produced at any one time must be very uniform in thickness, however, in order to obtain uniformly coloured chips. Slices with rough or torn surfaces lose excess solubles from ruptured cells and absorb larger amounts of fat.

It is necessary to remove the starch and other material released from the cut cells from the surface of slices so that the slices will separate readily and completely during frying. The slices are washed in stainless steel wire-mesh cylinders or drums rotating in a rectangular stainless steel tank. After washing and an additional rinse in similar equipment, the potatoes may or may not be dried.

Frying

The capacity of the fryer is generally the limiting factor in the process line. Most manufacturers currently use continuous fryers but some batch equipment is still employed.

Modern continuous fryers have the following essential elements: (1) a tank of hot oil in which the chips are cooked; (2) a means for heating and circulating the oil; (3) a filter for removing particles from oil; (4) a conveyor to carry chips out of the tank; (5) a reservoir in which oil is heated for adding to the circulating frying oil and (6) vapour-collecting hoods above the tank. Temperatures normally used are from 350 to 375° F at the receiving end and 320 to 345° F at the exit end.

The oil used for deep-fat frying of potato chips has two functions:

— it serves as a medium for transferring heat from a thermal source to the tuber slices;

— it becomes an ingredient of the finished product.

Use of highly refined oil is of great importance in flavour and stability of the crisps. Flavour, texture, and appearance are affected both by the amount of oil absorbed and its characteristics as it exists in the crisp. Oils change continuously during the frying process but the heat abuse resulting from the crisp cooking is relatively mild. Temperatures rarely rise above 385° F at any point. Better control over crisp colour could be obtained if the final stage of moisture removal could be achieved without the browning reaction that always accompanies it in the frying process.

Crisps may be sorted for size after frying, with the larger crisps being diverted to the bulk packs and larger pouches and the smaller pieces used for vending machine packs and other individual service containers. Potato crisp sizing is also accomplished by separating the peeled potatoes into large and small sizes, which are then sliced and fried separately. The crisps are salted immediately after they leave the fryer. It is important that the fat be liquid at this point to cause maximum adherence of the granules.

Powders containing barbecue spices, cheese, or other speciality materials may be added at this point. The salt may contain added enrichment materials or antioxidants. After salting, the crisps pass on to a conveyor belt where they are visually inspected and off-colour material is removed. The crisps are allowed to cool before packaging, better adherence of salt and flavour powders is obtained. Some consumers prefer the hard, curled-up crisp that is characteristic of the hand-kettle type of operation.

The special flavour of the hand-kettle crisp is said to be due, at least partly, to the starch retained on the cut surfaces of the potato slices as a result of the omission of a washing process after slicing. Starch-covered slices tend to stick together in the fryer so it is necessary to use devices to prevent clumping. The principal factors affecting potato crisp acceptability are piece size, colour, and of course, flavour. These factors are controllable primarily by selection of the raw material, adjustment of processing conditions, and packaging.

Storage Stability

If the frying oil is stabilised and has not deteriorated through use, and if the packaging is opaque and has a low moisture vapour transmittance rate (MVTR), a shelf-life of 4-6 weeks should be achieved when crisps are stored at temperatures of about 70° F. Once potato crisps are in the bag, the three forms of quality loss which have the greatest effect on consumer acceptance are breakage, absorption of moisture with loss of crispiness, and fat oxidation leading to development of rancid odours.

The mechanical abuse causing breaking of the crisps can be partially prevented by using stiff packaging material, making the package "plump" with contained air, and avoiding crushing in the shipping case. Absorption of moisture is prevented largely by proper choice of packaging material. Cellophane coated with various moisture barriers has proved to be a satisfactory pouch films for the relatively short shelf-life expected.

Light (especially fluorescent light) accelerates oxidation, so that opaque packaging material must be used to obtain maximum shelf-life. Potato crisps are considered commercially unacceptable when they have a moisture content above 3%, which is in equilibrium with a relative humidity of about 32%. The containers should have a high degree of resistance to moisture-vapour transfer. If pouches are used, foil-containing films are preferable, since they not only resist moisture-vapour transfer but reflect light.

Vegetable Juices

Vegetable juices are natural products constituted from cellular juice and a part of crushed pulp, from the tissues of some vegetables. These juices contain all valuable substances from the vegetables: vitamins, sugars, acids, mineral salts and pectic substances. The most important of these products is tomato juice; in a lower proportion there are also other juices.

Tomato Juice

This product is characterised not only by its organoleptical properties (taste, colour, flavour) but also by its vitamin content close to those of fresh tomatoes. Modem technology is oriented to a maximum maintenance of organoleptic properties and of vitamin content.

At same time, it is important to assure juice uniformity by avoiding cellulosic particle sedimentation. Juice stability is assured by a flash pasteurization which assures the destruction of natural micro-flora, while keeping the initial properties.

The modern technological flow-sheet covers the following main operations:

— *Pre Washing* is carried out by immersion in water, cold or heated up to 50° C (possibly with detergents to eliminate traces of pesticides). This operation is facilitated by bubbling compressed air in the immersion vessel/equipment.

— *Washing* is performed with water sprays, which in modern installations have a pressure of 15 at or more.

— *Sorting/control* on rolling sorting tables enables the removal of non-standard tomatoes—with green parts, yellow coloured, etc.

— *Crushing* in special equipment.

— *Preheating* at 55-60° C facilitates the extraction, dissolves pectic substances and contributes to the maintaining of vitamins and natural pigments. In some modern installations, this step is carried out under vacuum at 630-680 mm Hg and in very short time.

— *Extraction* of juice and part of pulp (maximum 80%) is performed in special equipment / tomato extractors with the care to avoid excessive air incorporation. In some installations, as an additional special care, a part of pulp is removed with continuous centrifugal separators.

— *De-aeration* under high vacuum of the juice brings about its boiling at 35-40° C.

— *Homogenisation* is done for mincing of pulp particles and is mandatory in order to avoid future potential product "separation" in two layers. Flash Pasteurisation at 130-150° C, time = 8-12 see, is followed by cooling at 90° C, which is also the filling temperature in receptacles (cans or bottles).

— Aspectic filling.

— *Closing of Receptacles* is followed by their inversion for about 5 to 7 minutes.

— *Cooling* has to be carried out intensely.

Full cans do not need further pasteurisation because the bacteria that have potentially contaminated the tomato juice during filling are easily destroyed at 90° C due to natural juice acidity. For bottles, it may be possible to avoid further sterilisation if the following conditions can be respected: washing and sterilising of receptacles, cap sterilisation, filling and capping under aseptic conditions, in a space with UV lamps. In so far as this is quite difficult to achieve it may be necessary to submit bottles to a pasteurisation in water baths.

The main characteristics of high quality tomato juice are:

— natural red colour;

— taste and flavour of fresh tomatoes;

— uniformity (without pulp sedimentation);

— total soluble solids: 6% minimum;

— total soluble substances (by refractometer): 5% minimum;

— vitamin C: 15 mg/100ml minimum.

In traditional processes it is recommended to:

— thoroughly wash and rinse the empty receptacles (including jar caps / covers and bottle crown corks) and then "sterilise" by keeping in boiling water for 30 mini

— add salt and lemon juice to the prepared receptacles just before filling;

— pasteurise closed glass receptacles (bottles or jars) according to conditions recommended in technological flow-sheets and which is summarised as follows:

Receptacle size	Pre-heating	Time of pasteurization
0.33 1	60° C	40 minutes
0.501	60° C	45 minutes
0.66 1	60° C	55 minutes
0.751	60° C	60 minutes
1.0 litre	60° C	70 minutes

Carrot Juice

This product represents an important dietetic product due to its high soluble pectin content. Technological flow-sheet is oriented to the maintaining of as high as possible a pectin content and covers the following steps:

- Pre-Washing
- Cleaning
- Washing
- Blanching in steam for 20 minutes
- Grating
- Pressing
- Juice In the pressed juice will then be incorporated 25% of grated carrot (non pressed)
- Homogenisation in colloidal mills
- Acidification with 0.25% citric or tartric acid
- De-Aeration
- Filling in receptacles (bottles or tinplate cans)
- Airtight Sealing
- Pasteurisation at 100° C for 30 minutes.

The main characteristics of a good quality carrot juice:

- uniformity (no separation in layers occurs during storage);
- good orange colour;
- pleasant taste, close to fresh carrot taste;
- total soluble solids: 12 %;
- total sugar content: 8%;
- beta-carotene: 1.3 mg/100 ml;
- soluble pectin: 0.4 %.

Red Beet Juice

The product is obtained following this technological flow-sheet: washing, cleaning, steam treatment / steaming (30-35 min at 1050 C), pressing, strain through small hole sieve, filling in receptacles, tight sealing / closing, sterilisation (25 min at 1 15° C). In order to improve taste, the juice is acidified with 0.3% citric or tartric acid.

Sauerkraut Juice

Sauerkraut juice is produced in some countries for its dietetic value (lactic acid and vitamin C content) and its refreshing taste. The juice which is the result of the fermentation of lactic acid from cabbage, mainly from sliced sauerkraut, is used.

The juice must be the result of a normal lactic fermentation, i.e. without butyric fermentation or other deterioration. A good quality juice must have an acidity of 1.4% lactic acid and a content of maximum 2.5% salt; this is obtained by the mixing of various sauerkraut qualities.

The collected juice (from sauerkraut production) is heated slightly in order to eliminate CO_2 gas and to obtain protein coagulation. Filtration of juice is the next technological step, followed by filling in receptacles, closing of receptacles and pasteurisation at 75-80° C for 4-5 minutes.

Concentrated Tomato Products

Tomato Paste

The product with highest production volumes among concentrated products is tomato paste which is manufactured in a various range of concentrations, up to 44% refractometric extract. Tomato paste is the product obtained by removal of peel and seeds from tomatoes, followed by concentration of juice by evaporation under vacuum.

In some cases, in order to prolong production period, it may be advisable or possible to preserve crushed tomatoes with sulphur dioxide as described under semiprocessed fruit "pulps". Technological flow-sheets run according to equipment/ installation lay-outs, which are especially designed for this finished product. Manufacturing steps fall into three successive categories:

- — obtaining juice from raw materials;
- — juice concentration and
- — tomato paste pasteurisation.

Obtaining juice from raw material. Preliminary operations (pre-washing, washing and sorting / control) are carried out in the same conditions as for manufacturing of "drinking" tomato juice described above. Next operation is removal of seeds from raw tomatoes: tomato crushing and seed separation with a centrifugal separator. Tomato pulp is pre-heated at 55-60° C and then passed to the equipment group for sieving: pulper, refiner and superrefiner with sieves of 1.5 mm, 0.8 mm and 0.4-0.5 mm respectively in order to give the smoothest possible consistency to the tomato paste.

Juice is concentrated by vacuum evaporation, a technological step which in modern installations runs continuously, tomato paste from the last evaporation step being at the specified concentration. In continuous installations with three evaporation steps (evaporating bodies), the juice is submitted in step / body I to pasteurisation at 85-900 C for 15 min and this will determine the microbiological stability of finished product. Vacuum degree corresponding to this temperature is 330 mm Hg.

In evaporating bodies II and III, temperatures are around 42-46° C and vacuum at 680700 mm Hg. Juice concentration occurs gradually and continuously in the three evaporating bodies. The advantages of continuous concentration are as follows:

- the taste, colour, flavour, "shine" and consistency of tomato paste are improved because:
  - the real concentration is performed in evaporating bodies II and III at low temperatures (42-46° C) and
  - the whole concentration process time from the input of juice in body I until the output of paste from body III is of about 1 hour (for paste with 30-35% refractometric extract).
- production capacity is raised by about 30% as compared to discontinuous installations with the same evaporation surface;
- the steam consumption is reduced by 60% because heating of bodies II and II is done with vapours resulting from juice evaporation in body I (double

effect); water and electricity consumptions are also reduced by 30-40 %.

Tomato paste pasteurisation assures the microbiological stability of the product. For this purpose, the paste coming out from concentration equipment is passed continuously and in a "forced" mode through a tubular pasteuriser from which it emerge at a temperature of 90-92° C.

Usual commercial tomato paste types are at concentrations of 24%, 28% and 32% refractometric extract. Sometimes it is possible to obtain a tomato paste with a concentration of 44% refractometric extract; for this purpose it is necessary to eliminate a part of cellulose from tomatoes, an operation performed in a separating turbine.

Tomato paste storage and preservation is carried out after packing which is done usually in drums, metallic cans or glass jars; some modern equipment has been developed for packing in aluminium bags. As far as the concentration of tomato paste is concerned it is not possible to reduce water content down to 30% which corresponds to a water activity aw of 0.700.75 (minimum limit of mould growing), it is necessary to take special measures.

Salt is not a preservative in itself but contributes to the lowering of water activity. In drums, the preservation of tomato paste with minimum 30% refractometric extract is carried out in two ways:

— the hot paste (about 90° C) flows directly from pasteurisation equipment into drums that have been previously steamed;

— the paste is cooled down to 30° C through a heat exchanger and is introduced into drums that have been previously steamed.

For preservation purposes, it is possible to add 3-8% salt.

Preservation with 3% salt must be carried out respecting the following criteria:

— processing of a healthy raw material;

— thorough washing and control;

— pasteurisation of concentrated paste and use of well prepared drums. Paste in drums has to be stored in cold storage rooms during the hot season.

Preservation in big metal cans of 5 and 10 kg capacity of tomato paste with a minimum of 30% refractometric extract can be achieved without sterilisation if the following conditions are respected:

— sterilisation by steam of cans and covers;

— filling of paste at 92-94° C;

— airtight sealing/ closing of cans;

— invert cans and then

— air cooling.

For small packages (tinplate cans of 1/10-1/1 or glass jars of same capacity) it is usual to use pasteurised paste, as hot as possible (92-94° C). The receptacles are first sterilised by steam. After airtight sealing, the receptacles are kept in boiling water for a short time in order to sterilise their inner surface and the paste in contact with inner receptacle surface. In some countries small receptacles are not further sterilised if the manufacturing is carried out in perfect hygienic and sanitary conditions.

Packing in small tinned aluminium tubes is carried out with concentrated paste, pasteurised and hot. Good quality tomato paste is an homogenous mass, with a high density, without foreign bodies (seeds, peel, etc.), with a red colour, and an agreeable taste and smell, close to those of fresh tomatoes. There are usually three types of tomato paste: 36, 30 and 24 which have refractometric extracts of respectively 34-38%, 28-32% and 24-26%. Paste of good quality must have a volatile acidity of maximum 0.15% as lactic acid. An 8% salt addition is accepted.

Concentrated Tomato Juice

Concentrated tomato juice is a product with 17-19 % refractometric extract and is a homogenous mass, finely sieved, without foreign bodies / and without any evidence of deterioration. A good quality product has a red colour, an agreeable and specific taste and smell. Modern technology uses the same installations, equipment and flow-sheets for concentrated tomato juice as for the production of tomato paste; the final concentration is thus regulated between the above specified limits.

The concentrated tomato juice is filled in receptacles (metal tinplate cans or glass bottles) and then pasteurised at 100° C during 15-25 minutes according to receptacle type. With modern production lines it should be possible to pass the concentrated tomato juice through a tubular pasteuriser and then pack aseptically and cool, without the need to pasteurise the receptacles.

Tomato Sauces

Under the USA Code of Federal Regulation 7 CFR 52, 1991 tomato sauce is the concentrated product prepared from the liquid extract from mature, sound, whole tomatoes, the sound residue from preparing such tomatoes for canning, or the residue from partial extraction of juice, or any combination of these ingredients, to which is added salt and spices and to which may be added one or more nutritive sweetening ingredients, a vinegar or vinegars, and onion, garlic, or other vegetable flavouring ingredients. The refractive index of the tomato sauce at 20° C is not < 1.3461.

These products are widespread in some countries and are used in order to spice some meals. Sauces can be obtained from fresh tomatoes or from concentrated products (tomato paste or concentrated tomato juice), those from fresh tomatoes being of superior quality.

Technological processing covers the following steps: concentrated juice processing, addition of flavour/taste ingredients (salt, sugar, vinegar, spices, etc.), boiling, fine sieving, filling of receptacles, closing and pasteurisation (45 min at 85° C). Tomato sauces which can be sweet, more or less spicy are prepared according to specific recipes.

Production Accidents and Product Defects

Tomato Juice

- "Separation" in layers is due to not enough homogenisation or low / insufficient viscosity. In the first case it is necessary to intensify homogenisation; and in second to increase the pre-heating temperature to 60° C in order to obtain protopectine hydrolisis and pectolitic enzymes inactivation.
- Moulding of the juice is brought about by significant infection of raw materials, inadequate washing and control or by use of contaminated packages. The preventive measures should be decided after cause analysis. Good pasteurisation can destroy all moulds but the bad juice taste remains.
- Fermentation of juice is manifested by a significant development of gases. Prevention methods are the same as for moulding.
- Tomato juice turns sour, without the formation of gases; this defect is initiated by thermophyl and thermoresistant bacteria; the juice acquires a vinegary taste. Prevention: maintenance of flash pasteurisation temperature at 130-135° C.
- Excessive vitamin C losses are due to a simultaneous action of heating and oxygen from air. Prevention:
  - prevent air going into crusher and extractor;
  - assure an intensive de-aeration (vacuum degree 700 mm Hg) at a temperature of at least 35-40° C; and
  - close receptacles in vacuum.
- Weak colour of tomato juice can be avoided by the utilisation of mature tomatoes and with a pulp of as red a colour as possible.

Tomato Paste and Concentrated Juice

- Presence of sand is caused by inadequate washing or by a significant contamination of raw material; this can be prevented by a more intensive pre-washing and washing of tomatoes.

— There may be mould especially at the surface of tomato paste packed in drums. Prevention:
  — accurate pre-washing and washing;
  — follow pasteurisation instructions;
  — pack in clean drums or receptacles; and
  — close receptacles immediately after filling.

— Fermentation is manifested by a weak alcohol smell or by a weak vinegar taste; when the fermentation is more advanced there is gas production in the product mass. Prevention: as for moulding prevention.

Tomato Sauces

Surface of the product turns black at the contact zone with air; this is due to the action of iron on the tannins from spices, tomato seeds, etc. Prevention:

— avoid iron equipment;
— avoid crushing of tomato seeds and
— seal receptacles in vacuum.

Pickles and Sauerkraut Technology

Natural Acidification Technology

Gherkins and Cucumbers

Raw materials must follow strict specifications for a high quality finished product; the following parameters must be considered as critical:

— adapt a uniform size according to the finished product requirements; for example, gherkins will need to have a maximum length of 9 cm for raw vegetables. Generally 15 cm size/length will be a maximum for high quality cucumber products in many countries. However, according to local preferences, bigger cucumbers could be also in demand.

— cylindrical or ovoidal shape;
— dark green colour;
— absence of surface defects due to cryptogamic diseases.

Cucumbers have to be picked at their ripeness for eating, when the sugar content is at about 1.5-2.2%, needed for lactic fermentation. Unripe cucumber does not have enough sugar.

The general technological flow-sheet is as follows:

— Reception
— Control
— Temporary Storage
— Grading By Size
— Washing
— Small Holes are made in large size cucumbers skin;
— Receptacle Filling: raw material is simply put in the receptacles in bulk, with care to arrange them in such a way that a maximum of pieces could be introduced;
— Salt Solution Preparation: 6% salt solution (NaCl);
— Salt Solution Addition: the salt solution is poured into the receptacle;
— Fermentation is carried out at 20-30° C, anaerobically. This step takes generally 4-8 weeks. Acidity reaches a value up to 1.5% lactic acid (and in some exceptional cases up to 2% lactic acid) which corresponds to a maximum pH value of 4.1.
— Storage; after the last fermentation stage, drums and other receptacles have to be stored at low temperature; best conditions for 12 months shelf life should be below + 15° C. Storage temperature will determine the shelf life of the products.
— Addition of 1000 ppm potassium sorbate will prevent mould development without having any influence on lactic fermentation.
— Raw material grading by size is a very important technological step. In order to accelerate brine

penetration, mainly for medium to large size cucumbers, the practice of making small holes in the raw material skins is generally recommended.

— A major factor influencing the quality of lactic fermented cucumbers is the water durity; optimal results are obtained at 15-20° durity.

Cucumber consistency / texture is influenced by the formation of calcium pectate with the pectic substances from raw material tissues. In some countries, calcium chloride (0.3-0.5 %) is added in order to firm up the cucumber consistency. Chlorinated water which still contains active chlorine can inhibit or even stop the lactic fermentation.

Sauerkraut

In some countries cabbages are submitted to lactic fermentation as whole vegetables; however, in many countries the cabbage is shredded before fermentation. As shredded cabbage and its technology is at the basis of an important industry, giving good quality products, with a uniform fermented product and with good keeping quality and ease of distribution, this will be described first.

Cabbage as raw material for sauerkraut must be sound, ripe for eating, well-leafed and from suitable varieties. Optimum total sugar level needed for the lactic fermentation is 24%; generally good quality raw material contains up to 30-60 mg/100 g of vitamin C.

Shredded Sauerkraut

The technological flow sheet is as follows:

— Reception

— Control

— Temporary Storage is carried out in bulk, up to a height of about 1 m, during few days. This step produces a heat generation which facilitates later fermentation by the softening of tissues.

— Removal of External Leaves

Coring is done with a specially adapted mechanical screw; this operation generates small particles of finely divided cabbage which will be mixed with the main part of vegetable during shredding / chopping. The core represents about 10% from the whole cabbage, is rich in sugar and vitamin C, but being too high in fibre content needs to be chopped separately as described. Shredding/Cutting of cabbage is carried out with complex specific equipment which is generally installed directly on the "top" of fermentation silos and is mobile, installed on rails and moves all along the silos. The dimension of resulting shredded cabbage is about 2-3 mm thick.

The same complex equipment is designed to grind the added salt to fine particles and to distribute shredded cabbage and ground salt in an uniform manner to the fermentation silos. The usual capacity of fermentation silos is up to 30 tons, with separate compartments of 45 tons each. Salt Addition is carried out by the equipment described above; the proportion of salt is 2-2.5% with respect to cabbage. This proportion must not be changed because the salt in this technology does not have a preservative role but only that to extract from cabbage the juice needed for fermentation.

It is be preferable to obtain a fairly light pressure on cabbage just after salt addition with some simple mechanical means. This is important in order to:

— create an anaerobic medium for fermentation;

— facilitate external diffusion of cellular juice;

— assure a rational use of the fermentation space.

Fermentation: The maximum acidity level obtained is generally of about 1.5% lactic acid (and very rarely 2.5%); this is obtained in 4-6 weeks. Optimal acidity is 1.0-1.8% and pH value 4.1 or lower. Fermentation temperature is at 20-25° C in the first phase and needs to be lowered then to 14-18° C. During fermentation, the brine from each storage / fermentation silo cell is periodically circulated with a pump in order to uniformise the fermentation process.

Storage is performed in same silos used for fermentation, or the finished products is removed from silos and packed in drums and other receptacles according to distribution schedule. These silos are usually made of reinforced concrete and coated with gritstone plates or with an acid-resisting material layer. At small scale and in traditional processing, shredded sauerkraut can be obtained by using simple available glass or rigid plastic receptacles. At home, this process can use glass jars and / or local / traditional pottery receptacles from a minimum size of 2-3 kg up to the available / practical sizes (better limited to 10-15 kg). In some countries shredded sauerkraut is preserved in receptacles by pasteurization, once the fermentation process has been completed.

Whole Sauerkraut

According to the consumer preference in different countries and to the specific situations it is also usual to preserve whole cabbages by lactic fermentation. At small or medium scale operations, whole cabbage could be processed/ fermented in cylindric receptacles like 30 to 200 litre rigid plastic drums, or rectangular receptacles made from food grade rigid plastic. It is possible to find this type of drum in a significant number of developing countries. These two types of rigid plastic receptacles could also be used for shredded sauerkraut production.

Prepared whole cabbages are put into fermentation receptacles and a 5-6 % salt concentration brine is poured on top. The fermentation conditions are the same as for shredded sauerkraut. In order to assure a uniform fermentation and to avoid a strict anaerobic (butyric) fermentation it is necessary to apply a periodic juice "aeration".

Other Acidified Vegetables

In principle all vegetables with a sugar content of at least 2 % could be preserved by lactic fermentation. From a practical point of view it is mainly the following vegetables which are preserved by this technology: unripe tomatoes (green tomatoes), peppers, eggplant, carrots and cauliflower, alone or usually in a mix with cucumber as mixed pickles.

Fermentation of individual vegetables is carried out according to a flow-sheet as described for whole sauerkraut. The type of cut, brine concentration and frequency of operating steps have to be adapted to each case; green tomatoes are fermented as whole vegetable.

Artificial Vegetable Acidification Technology

This technology is based on the addition of food grade vinegar which has a bacteriostatic action in concentrations up to 4 % acetic acid and bactericidal action in higher concentrations. Vegetables preserved in vinegar need to reach, after equilibrium between vinegar and water contained in vegetables, a final concentration of 2-3 % acetic acid in order to assure their preservation. To achieve this final concentration, a 6-9 % acetic acid vinegar is used, as related to the specific ratios vinegar/ vegetables. In vinegar pickles, salt (2-3 %) and sometimes sugar (2-5 %) are also added. If the vinegar concentration is lower than 2%, vinegar pickles need to be submitted to a pasteurisation in order to assure their preservation.

Cucumbers in Vinegar

This represents the basic product obtained by this technology. Cucumbers have to be wholesome, with a soft texture and not have reached eating maturity. They must have a low sugar content because in this technology there is no lactic fermentation involved. Dimensions are up to 12 cm length, with a preference for small cucumbers.

The technological steps are the followings:

- Size Grading
- Washing
- Arrange In Receptacles—glass jars, etc.
- Pouring Of Vinegar is usually carried out at room temperature; however, hot vinegar addition enables a sterilisation of cucumber surface and facilitates vinegar penetration in vegetable tissues.
- Salt (Sugar) Addition

— Spicing Addition
— The technological cycle of artificial acidification is considered completed when acetic acid concentration reaches an equilibrium value; the time needed is about 2 weeks.

When equilibrium concentration in acetic acid is below 2 %, the cucumbers are submitted to a pasteurisation for 20 min at 90-1000 C in order to assure their preservation. Cucumbers in vinegar with previous lactic fermentation are excellent quality products because the lactic fermentation improves the taste of these cucumbers. The principle of this process is to assure preservation both by acetic acid and by lactic acid simultaneously.

Technological processing flow-sheet is as follows: small cucumbers ("cornichons" or "gherkins") are washed, brushed and small holes are made in the skin; the vegetables then are put in drums with slightly warm 6% brine which also contains spices. The lactic fermentation runs for few days up to a lactic acid concentration of about 0.5 %. The cucumbers are removed from the brine, washed thoroughly and well drained.

Preservation is usually done in glass jars by pouring a normally flavoured vinegar with about 9% acetic acid usually in order to bring the final concentration to 3% calculated as acetic acid. In order to obtain a high quality product only wine vinegar should be used. In some pickles (e.g. in "Cornichons") the usual level of wine vinegar is set at 20 % of packaged product total weight; some alcohol vinegar could be still added and final concentration will be adjusted as described above.

Other Vinegar Pickles

One type in this category is represented by other vegetables acidified with vinegar separately or in a mix (red peppers, sweet green pepper, green tomatoes, cauliflower, etc.). The preparation steps are similar to the ones used for cucumbers in vinegar. Significant quantities of special mixed vegetables in vinegar are manufactured in many countries, with the international name of "mixed pickles" with following composition: small cucumbers ("cornichons"/"gherkins")— maximum 70 mm in length -, sliced carrots, cauliflower, small onions (less than 25 mm diameter), mushrooms etc. and spices.

The vegetables are acidified separately in vinegar and then are put into receptacles (glass jars); a flavoured vinegar, salted and sweetened with acetic acid concentration of 3-5% is poured over them. In the case of lower acetic acid concentrations, a pasteurisation at 90° C for 10-20 minutes is applied according to the receptacle size.

Canned Vegetable

Canned vegetables can be classified as follows:

- — canned products in salt brine;
- — canned products in tomato concentrated juice;
- — canned products in vegetable oil.

Canned Vegetables in Tomato Juice

General technological flow-sheet covers two types of operations:

- — Preparation of vegetables is similar to the one described for canned vegetables in salt brine: sorting, washing, grading, cutting, blanching and cooling; the exception is for spices which are not blanched.
- — Preparation of canned products covering: receptacles filling with vegetables, adding concentrated tomato juice (with minimum 8% refractometric extract), hermetic closing/sealing of receptacles, sterilisation and cooling of receptacles.

One usual composition for mixed vegetables in tomato juice is:

- — eggplants (slices): 20%
- — peppers (cut): 20%
- — carrots (slices): 15%
- — green peas: 5%

— green beans (pods): 18%
— okra (whole): 8%
— tomatoes (whole or halves): 14%

RE = Refractometric extract

Each vegetable is prepared separately as in general canning operation description. At receptacle filling for mixed vegetables products, each vegetable should be introduced separately in specified proportions; hot concentrated tomato juice (at least 700 C) is poured onto the vegetables. Sterilisation is carried out according to the instructions given in Table 9.7.2 and then receptacles have to be thoroughly cooled.

Heat Preservation Operations Canning

The success of heat preservation operations lies in:

— selecting suitable fruit and vegetables in good conditions;
— preparing them hygienically and skilfully;
— packing them in cans which are hermetically sealed and then processed under fixed conditions of time and temperature;
— cooling these cans carefully and storing them under conditions which will not cause deterioration of either the cans or their contents.

Selection of Raw Materials

It is appreciated that some varieties of fruit and vegetables are not suitable for canning, either because they are uneconomical to prepare or because the colour, flavour or texture are poor. Suitable varieties must be available to the canner in quantities sufficient to meet his requirements and in sound conditions for canning.

The flow to the cannery should be regulated in order that perishable materials are not left for a long time before being handled, since any delay will cause deterioration. Apart from the main ingredients, be it fruit or vegetables, minor ingredients also require careful selection. Sugar, salt, water and spices for instance may all be contaminated with spoilage organisms, so constant testing of all raw materials is essential.

Preparation

This is carried out by various methods, including grading, trimming, peeling, washing and blanching. All equipment must be scrupulously clean and preparation should be completed quickly and carefully in order to keep the bacterial load as low as possible. Thorough washing of vegetable is necessary to remove spores of heat resistant bacteria which are present in large numbers in the soil. Blanching in steam or hot water is of no avail against these heat resistant spores because of the comparatively low temperatures involved.

Reasons for blanching are:

- the removal of gas from the tissues of the raw material;
- the shrinkage of this material;
- the inhibition of enzymic reactions, which, if not checked, will adversely affect the colour and nutritive value of the food.

Filling

Filling, be it mechanical or by hand, requires careful attention. The cans must be clean and the correct weight of foodstuffs must be added. Under-filled cans will be underweight and the headspace will be too large, resulting in too much air being left in the can. Overfilling may lead to seams being strained during processing and to ends becoming distorted and bulged. If the product forms hydrogen on storage as is the case with coloured fruits, swelling of the can due to hydrogen pressure will occur more quickly in an overfilled can than in one which has been correctly filled. Overfilling also affects heat penetration in the can and may lead to spoilage outbreaks.

Air Removal

Before the can is seamed, air must be removed from the contents and the headspace. Normally, this is carried out by passing the cans through a steam box until the temperature at the centre of the can is at least 160° F. This operation, termed exhausting, is necessary for the following reasons:

- to minimise strains on the seams due to expansion of air during the processing period;
- to remove oxygen which accelerates corrosion in the can and also causes oxidation of the food with possible serious effects on colour and flavour;
- to reduce the destruction of vitamin C;
- to enable a vacuum to be formed when the can is cooled.

This ensures that the ends remain concave, even when storage temperatures are a little higher than usual, and also acts as a reservoir for hydrogen which may be formed by reactions between the can and its contents. Thus a high vacuum makes for a long shelf life. Large cans, however, should not reach such a high exhaust temperature before seaming as smaller cans because of the danger of the can body collapsing on cooling, a condition known as "panelling".

Double Seaming

The can should be double-seamed as soon as the correct centre temperature has been attained. Any delay between exhausting and seaming will lead to loss of vacuum and may lead to bacterial spoilage. The quality of the double seam must, of curse, be frequently checked.

Heat Processing

After seaming, the cans are heated for a definite time at a definite temperature to kill or inhibit organisms which may cause spoilage. This operation is termed "heat processing". The times and temperatures required for "heat processing" of various packs have been determined experimentally to ensure that spores of the most heat resistant food poisoning organisms known, Clostridium botulinum, are destroyed. There are other organisms, however, whose spores are more heat resistant than those of Clostridium botulinum and which although they will not cause food poisoning may cause spoilage and for this reason the minimum heat processing time is often exceeded by recommendations made by laboratories.

At the same time there is a limit to the amount of heating which a canned food may be given without spoiling its flavour, texture and colour and this also has to be taken into consideration when process recommendations are made. Bacterial spores have a greater resistance to heat when the growth-medium is neutral or near neutral, and neutrality is normally required for bacterial growth to commence. Because of this, canned foods have been broadly divided into two groups:

— "acid" foods having a pH of 4.5 or lower and

— "non-acid" foods having a pH of more than 4.5.

"Non-acid" foods (vegetables) must, therefore be "heat processed" at high temperatures using steam under pressure, whereas "acid" foods (fruit) may be processed at the (lower) temperature of boiling water, since this will kill moulds and yeasts and if any bacterial spores survive the combination of acid and heat, they will be inhibited from growth by the acid environment. The rate of destruction by heat follows a definite pattern, the same proportion of the surviving bacteria being destroyed in successive units of time.

The more bacteria there are in a pack, the more time will be need to reduce their numbers. For this reason, it is essential that the initial number of bacteria be kept low, and this may be achieved by ensuring fast and hygienic handling at all stages in the cannery. Pressure gauges and retort temperature control equipment must be checked frequently for accuracy. Processing times and temperatures must be strictly adhered to, and complete removal of air from the retort during processing must be achieved by adequate venting. Failure to remove the air completely will result in their being cold spots in the retort and intermittent spoilage is likely.

Cooling

As soon as the heat processing time is completed, the cans are cooled in chlorinated water as rapidly as possible without damaging them. Cans processed in steam develop high internal pressure because of the expansion of the foodstuff, the expansion of air in the can and the increase in the vapour pressure of the water in the can.

During the heat process, these pressures are counter-balanced to some extent by the pressure of the steam in the retort, but on releasing this steam pressure at the commencement of the cooling period, the pressure in the can may be sufficient to strain the seams seriously and may even distort the ends. Cans of A21/2 size or larger, when processed at temperatures of 240° F or more, are liable to undergo permanent distortion, such as peaking.

This may be avoided by pressure cooling, which involves replacing steam pressure by air pressure before introducing water to the retort, and maintaining this until the pressure inside the can has fallen to a safe level. This presents difficulties, since if the air pressure is maintained after the can has developed a vacuum, the can body is liable to collapse. Where pressure-cooling is not carried out, the retort pressure is allowed to drop slowly to atmospheric pressure and the cans are then cooled with water.

Storage

After cooling, the cans should be stored in cool, dry conditions. The maintenance of a constant temperature is desirable, since a rise in temperature may lead to condensation of moisture on the can, with possible rusting. Cool conditions are required because storage at higher temperatures not only causes chemical and physical changes in the product and the container but also introduces a risk of thermophilic spoilage.

Other known causes of container spoilage in storage are the use of labels and cardboard cases which have too high a chloride content, and the use of unseasoned wood in the manufacture of packing cases, all of which tend to cause rust formation on the cans.

References

Beychok, Milton R., *Aqueous Wastes from Petroleum and Petrochemical Plants*, 1st Edition, John Wiley & Sons, 1967.

Coulson, J. M. and Richardson, J. F., *Chemical Engineering*, 3rd Edition, Vols. 1, 2, Pergamon, Oxford, 1977, 1978.

Emken, E. A., "Nutrition and biochemistry of trans and positional fatty acid isomers in hydrogenated oils," *Annual Reviews of Nutrition*, 4: 339-376, 1984.

Toledo, R.T., *Fundamentals of Food Process Engineering*, 2nd Edition, van Nostrand, Reinhold, New York, 1991.

6

Coffee Processing

Quality coffees must be picked by hand, a process that takes from three to four visits per tree each year. This is because coffee cherries do not ripen at the same time. A branch of a tree might simultaneously bear blossoms, green fruit, and ripe cherries. A good picker can pick about 200 pounds of coffee cherries in one day. This equals about 50 pounds of green coffee beans or 39 pounds of roasted coffee. Once the coffee cherries have been picked, the beans must be removed from them.

Coffee Harvesting Times

Each year coffee is harvested during the dry season when the coffee cherries are bright red, glossy, and firm. Ripe cherries are either harvested by hand, stripped from the tree with both unripe and overripe beans, or all the coffee beans are collected using a harvesting machine. These processes are called selective picking, stripping, and mechanical harvesting, respectively. To maximise the amount of ripe coffee harvested, it is necessary to selectively pick the ripe coffee beans from the tree by hand and leave behind unripe, green beans to be harvested at a later time.

Brazil's Process of Coffee Bean Harvesting

In Brazil, harvesting the same coffee tree several times is less cost effective than separating and discarding the unripe or overripe cherries. Therefore, Brazil typically harvests using the stripping method when 75% of the coffee crop is perfectly ripe. Stripping is feasible and cost effective in Brazil due to the uniform maturation of Brazilian coffees. In stripping, the coffee beans are pulled from the tree and fall to the ground where they are caught by sheets. The beans are removed from tree debris by tossing the coffee in the air allowing the wind to carry away sticks and leaves.

The coffee is then put in 60 L green baskets, which is the tool are measurement used by coffee producers to determine wages. Some coffee estates, such as Fazenda Monte Alegre in Sul de Minas Brazil, have a computerised system to determine wages for picking coffee beans. This system accounts for the amount of coffee collected from each person, the difficulty of the coffee harvesting conditions, and the production of the region being harvested.

Picking

A coffee plant usually starts to produce flowers 3-4 years after it is planted, and it is from these flowers that the fruits of the plant (commonly known as coffee cherries) appear, with the first useful harvest possible around 5 years after planting. The cherries ripen around eight months after the emergence of the flower, by changing colour from green to red, and it is at this time that they should be harvested. In most coffee-growing countries, there is one major harvest a year; though in countries like Colombia, where there are two flowerings a year, there is a main and secondary crop.

In most countries, the coffee crop is picked by hand, a labour-intensive and difficult process, though in places like Brazil, where the landscape is relatively flat and the coffee fields immense, the process has been mechanized. Whether picked by hand or by machine, all coffee is harvested in one of two ways:

Strip Picked

The entire crop is harvested at one time. This can either be done by machine or by hand. In either case, all of the cherries are stripped off of the branch at one time.

Selectively Picked

Only the ripe cherries are harvested and they are picked individually by hand. Pickers rotate among the trees every 8 - 10 days, choosing only the cherries which are at the peak of ripeness. Because this kind of harvest is labour intensive, and thus more costly, it is used primarily to harvest the finer arabica beans. The labourers who pick coffee by hand receive payment by the basketful. As of 2003, payment per basket is between US$2.00 to $10 with the overwhelming majority of the labourers receiving payment at the lower end. An experienced coffee picker can collect up to 6-7 baskets a day. Depending on the grower, coffee pickers are sometimes specifically instructed to not pick green coffee berries since the seeds in the berries are not fully formed or mature.

This discernment typically only occurs with growers who harvest for higher end/speciality coffee where the pickers are paid better for their labour. Mixes of green and red berries, or just green berries, are used to produce cheaper mass consumer coffee beans, which are characterised by a displeasingly bitter/astringent flavour and a sharp odour. Red berries, with their higher aromatic oil and lower organic acid content, are more fragrant, smooth, and mellow. As such coffee picking is one of the most important stages in coffee production, and is the chief determinant for the quality of the end product.

Processing Method

Wet Process

In the Wet Process, the fruit covering the seeds/beans is removed before they are dried. Coffee processed by the wet method is called wet processed or washed coffee. The wet method requires the use of specific equipment and substantial quantities of water. After the Green coffee is picked the coffee is sorted by immersion in water. Bad or unripe fruit will float and the good ripe fruit will sink. The skin of the cherry and some of the pulp is removed by pressing the fruit by machine in water through a screen. The bean will still have a significant amount of the pulp clinging to it that needs to be removed. This is done either by the classic ferment-and-wash method or a newer procedure variously called machine-assisted wet processing, aquapulping or mechanical demucilaging:

Ferment-and-Wash Method

In the ferment and wash method of wet processing the remainder of the pulp is removed by breaking down the cellulose by fermenting the beans with microbes and then washing them with large amounts of water. Fermentation can be done with extra water or, in "Dry Fermentation", in the fruit's own juices only. The fermentation process has to be carefully monitored to ensure that the coffee doesn't acquire undesirable, sour flavours. For most coffees, mucilage removal through fermentation takes between 24 and 36 hours, depending on the temperature, thickness of the mucilage layer and concentration of the enzymes.

The end of the fermentation is assessed by feel, as the parchment surrounding the beans loses its slimy texture and acquires a rougher "pebbly" feel. When the fermentation is complete, the coffee is thoroughly washed with clean water in tanks or in special washing machines.

Machine-assisted Wet Processing

In machine-assisted wet processing, fermentation is not used to separate the bean from the remainder of the pulp; rather, this is done through mechanical scrubbing. This process can cut down on water use and pollution since ferment and wash water stinks. In addition, removing mucilage by machine is easier and more predictable than removing it by fermenting and washing. However, by eliminating the fermentation step and prematurely separating fruit and bean, mechanical demucilaging can remove an important tool that mill operators have of influencing coffee flavour.

Furthermore, the ecological criticism of the ferment-and-wash method increasingly has become moot, since a combination of low-water equipment plus settling tanks allows conscientious mill operators to carry out fermentation with limited pollution. After the pulp has been removed what is left is the bean surrounded by two additional layers, the silver skin and the parchment. The beans must be dried to a water content of about 10% before they are stable. Coffee beans can be dried in the sun or by machine but in most cases it is dried in the sun to 12-13% moisture and brought down to 10% by machine. Drying entirely by machine is normally only done where space is at a premium or the humidity is too high for the beans to dry before mildewing.

When dried in the sun coffee is most often spread out in rows on large patios where it needs to be raked every six hours to promote even drying and prevent the growth of mildew. Some coffee is dried on large raised tables where the coffee is turned by hand. Drying coffee this way has the advantage of allowing air to circulate better around the beans promoting more even drying but increases cost and labour significantly.

After the drying process (in the sun and/or through machines), the parchment skin or pergamino is thoroughly dry and crumbly, and easily removed in the Hulling process. Coffee occasionally is sold and shipped in parchment or en pergamino, but most often a machine called a huller is used to crunch off the parchment skin before the beans are shipped. Any wet processing of coffee produces coffee wastewater which can be a pollutant. Around 130 liters of fresh water is required to process one kilogram of quality coffee.

Coffee Bean Moisture Measurement

Before shipment, coffee is dried and a coffee moisture meter is used to measure coffee bean moisture. Coffee must be dried from approximately 60% moisture content to 11-12% moisture content. Coffee is typically dried on large patios made of asphalt or cement and then transferred to mechanical dryers. The coffee on the drying patios is shifted every 30-40 minutes and is shaped into long rows of no more than 5 cm in height. Next to each row is open ground, which is warmed and dried by the sun.

The coffee is then shifted onto the dry portion of the patio. This helps accelerate the coffee drying process and prevents fermentation and moldy beans from developing. This method is widely used in Brazil, but less widely used in Guatemala or Costa Rica where the coffee is more often piled perpendicularly to the old piles. Drying coffee solely by patio takes 6-7 days for washed coffees, 8-9 days for pulped naturals (semi-washed), and 12-14 days for natural (dry-processed) coffees.

This is why coffee beans are typically dried on a patio until they reach a moisture content of 15% and are then transferred to mechanical dryers. Once the coffee reaches a 25% moisture content or less, it can be piled at night and covered with cotton cloths to allow the coffee to breath. If it rains, these piles can also be covered with plastic. Coffee should not be covered with burlap sacks since this will impart a distinct burlap flavour and aroma to the coffee.

Coffee Drying Stages

In a study done in Kenya, Kamau reports that there are six stages to drying coffee.

- Skin drying. Moisture 55-45%.
- White Stage drying. Moisture 44-33%.
- Soft Black stage. Moisture 32-22%.
- Medium Black Stage. Moisture 21-16%
- Hard Black Stage. Moisture 15-12%
- Fully dry coffee and conditioning. 11-10%.

Coffee Drying Equipment

There are several coffee dryer systems available. Many older dryers are converted grain dryers that are not as efficient as the new horizontal barrel dryers. The new coffee dryers are designed to mix the coffee evenly to ensure uniform drying. Drying coffee by using mechanical dryers accelerates the slowest part of the coffee drying process (15-11%) and helps prevent fermentation.

In some environments that have a high humidity the entire drying process must take place in mechanical dryers.

Mechanical dryers should never be set higher than 40-45°C and this question should be asked before buying any coffees that have been mechanically dried. At higher temperatures the germ is killed and the flavour potential of the coffee is ruined. At extreme temperatures the bean crystallizes, and when smashed with a hammer, will break like glass.

The best, but least utilised method of drying coffee is by using drying tables. In this method the pulped and fermented coffee is spread thinly on raised beds, which allows the air to pass on all sides of the coffee. The coffee is mixed by hand and the drying that takes place is more uniform and fermentation is less likely. Most coffee from Africa is dried in this manner and select coffee farms around the world are following their lead.

Method of Processing Coffee

Dry process, also known as unwashed or natural coffee, is the oldest method of processing coffee. The entire cherry after harvest is first cleaned and then placed in the sun to dry on tables or in thin layers on patios:

Cleaning

The harvested cherries are usually sorted and cleaned, to separate the unripe, overripe and damaged cherries and to remove dirt, soil, twigs and leaves. This can be done by winnowing, which is commonly done by hand, using a large sieve. Any unwanted cherries or other material not winnowed away can be picked out from the top of the sieve. The ripe cherries can also be separated by flotation in washing channels close to the drying areas.

Drying

The coffee cherries are spread out in the sun, either on large concrete or brick patios or on matting raised to waist height on trestles. As the cherries dry, they are raked or turned by hand to ensure even drying and prevent mildew. It may take up to 4 weeks before the cherries are dried to the optimum moisture content, depending on the weather conditions. On larger plantations, machine-drying is sometimes used to speed up the process after the coffee has been pre-dried in the sun for a few days.

The drying operation is the most important stage of the process, since it affects the final quality of the green coffee. A coffee that has been overdried will become brittle and produce too many broken beans during hulling (broken beans are considered defective beans). Coffee that has not been dried sufficiently will be too moist and prone to rapid deterioration caused by the attack of fungi and bacteria.

The dried cherries are stored in bulk in special silos until they are sent to the mill where hulling, sorting, grading and bagging take place. All the outer layers of the dried cherry are removed in one step by the hulling machine. The dry method is used for about 95% of the Arabica coffee produced in Brazil, most of the coffees produced in Ethiopia, Haiti and Paraguay, as well as for some Arabicas produced in India and Ecuador. Almost all Robustas are processed by this method. It is not practical in very rainy regions, where the humidity of the atmosphere is too high or where it rains frequently during harvesting.

Semi Dry Process

Semi dry is a hybrid process in very limited use in Brazil and Sumatara/Sulawesi. The cherry is passed through a screen to remove the skin and some of the pulp like in the wet process but result is dried in the sun and not fermented or scrubbed. Molecular density sorting is one of the best ways to separate broken, small, undeveloped, and otherwise defective coffee beans. There are usually 2-3 stages of density sorting. In the first stage, the very dense rocks and stones are removed from the coffee. In the second stage, the coffee is hulled and the debris is removed. The third stage is most important and uses a densimetric table to separate the coffee into three or more densities. The settings on the machine must be continuously monitored to ensure that proper density sorting is occurring.

Milling

The final steps in coffee processing involve removing the last layers of dry skin and remaining fruit residue from the now dry coffee, and cleaning and sorting it.

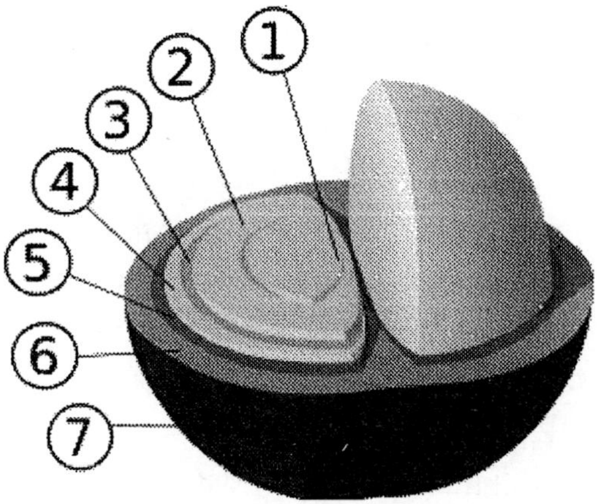

Figure 1. Structure of coffee berry and beans: 1: center cut 2:bean (endosperm) 3: silver skin (testa, epidermis), 4: parchment (hull, endocarp) 5: pectin layer 6: pulp (mesocarp) 7: outer skin (pericarp, exocarp)

These steps are often called dry milling to distinguish them from the steps that take place before drying, which collectively are called wet milling.

Hulling

The first step in dry milling is the removal of what is left of the fruit from the bean, whether it is the crumbly parchment skin of wet-processed coffee, the parchment skin and dried mucilage of semi-dry-processed coffee, or the entire dry, leathery fruit covering of the dry-processed coffee. Hulling is done with the help of machines, which can range from simple millstones to sophisticated machines that gently whack at the coffee.

Polishing

This is an optional process in which any silver skin that remains on the beans after hulling is removed in a polishing machine. This is done to improve the appearance of green coffee beans and eliminate a byproduct of roasting called chaff. It is decried by some to be detrimental to the taste by raising the temperature of the bean through friction which changes the chemical makeup of the bean.

Cleaning and Sorting

Sorting by Size and Density: Most fine coffee goes through a battery of machines that sort the coffee by density of bean and by bean size, all the while removing sticks, rocks, nails, and miscellaneous debris that may have become mixed with the coffee during drying. First machines blow the beans into the air; those that fall into bins closest to the air source are heaviest and biggest; the lightest (and likely defective) beans plus chaff are blown in the farthest bin. Other machines shake the beans through a series of sieves, sorting them by size. Finally, a machine called a gravity separator shakes the sized beans on a tilted table, so that the heaviest, densest and best vibrate to one side of the pulsating table, and the lightest to the other.

Sorting by Color: The final step in the cleaning and sorting procedure is called color sorting, or separating defective beans from sound beans on the basis of color rather than density or size. Color sorting is the trickiest and perhaps most important of all the steps in sorting and cleaning. With most high-quality coffees color sorting is done in the simplest possible way: by hand. Teams of workers pick discolored and other defective beans from the sounds beans. The very best coffees may be hand-cleaned twice (double picked) or even three times (triple picked). Coffee that has been cleaned by hand is usually called European preparation; most speciality coffees have been cleaned and sorted in this way.

Color sorting can also be done by machines. Streams of beans fall rapidly, one at a time, past sensors that are set according to parameters that identify defective beans by value (dark to light) or by color. A tiny, decisive puff of compressed air pops each defective bean out of the stream of sound beans the instant the machine detects an anomaly. However, these machines are currently not used widely in the coffee industry for two reasons. First, the capital investment to install these delicate machines and the technical support to maintain them is daunting. Second, sorting coffee by hand supplies much-needed work for the small rural communities that often cluster around coffee mills. Nevertheless, computerised color sorters are essential to coffee industries in regions with relatively high standards of living and high wage demands.

Grading is the process of categorising coffee beans on the basis of various criteria such as size of the bean, where and at what altitude it was grown, how. it was prepared and picked, and how good it tastes, or its cup quality. Coffees also may be graded by the number of imperfections (defective and broken beans, pebbles, sticks, etc.) per sample. For the finest coffees, origin of the beans (farm or estate, region, cooperative) is especially important. Growers of premium estate or cooperative coffees may impose a level of quality control that goes well beyond conventionally defined grading criteria, because they want their coffee to command the higher price that goes with recognition and consistent quality.

All coffee, when it was introduced in Europe, came from the port of Mocha in what is now modern day Yemen. To import the beans to Europe the coffee was on boats for a long sea voyage around the Horn of Africa. This long journey and the exposure to the sea air changed the coffee's flavour. Later, coffee spread to India and Indonesia but still required a long sea voyage.

Once the Suez Canal was opened the travel time to Europe was greatly reduced and coffee whose flavour had not changed due to a long sea voyage began arriving. To some degree, this fresher coffee was rejected because Europeans had developed a taste for the changes that were brought on by the long sea voyage. To meet this desire, some coffee was aged in large open-sided warehouses at port for six or more months in an attempt to simulate the effects of a long sea voyage before it was shipped to Europe.

Although it is still widely debated, certain types of green coffee are believed to improve with age; especially those that are valued for their low acidity, such as coffees from Indonesia or India. Several of these coffee producers sell coffee beans that have been aged for as long as 3 years, with some as long as 8 years. However, most coffee experts agree that a green coffee peaks in flavour and freshness within one year of harvest, because over-aged coffee beans will lose much of their essential oil content.

Decaffeination Process

Decaffeination is the process of extracting caffeine from green coffee beans prior to roasting. The most common decaffeination process used in the United States is supercritical carbon dioxide (CO_2) extraction. In this process, moistened green coffee beans are contacted with large quantities of supercritical CO_2 (CO_2 maintained at a pressure of about 4,000 pounds force per square inch (28 MPa) and temperatures between 90 and 100 °C [194 and 212 °F]), which removes about 97 % of the caffeine from the beans.

The caffeine is then recovered from the CO_2, typically using an activated carbon adsorption system. Another commonly used method is solvent extraction, typically using oil (extracted from roasted coffee) or ethyl acetate as a solvent. In this process, solvent is added to moistened green coffee beans to extract most of the caffeine from the beans. After the beans are removed from the solvent, they are steam-stripped to remove any residual solvent.

The caffeine is then recovered from the solvent, and the solvent is re-used. Water extraction is also used for decaffeination. Decaffeinated coffeebeans have a residual caffeine content of about 0.1 % on a dry basis. Not all facilities have decaffeination operations, and decaffeinated green coffee beans are purchased by many facilities that produce decaffeinated coffee.

Storing Coffee Beans

Coffee must be stored in dry and cool conditions. Exposure to the sun or moisture will rapidly deteriorate the coffee. Storage in pergamino until right before the shipping time will help preserve the coffee. Burlap bas are often used for coffee bean storage because they allow air flow. They also preserve the coffee longer than plastic or paper bags. Burlap bags should be aired on the patios before storing coffee to prevent a baggy flavour or burlap scent from being imparted to the coffee.

Research in Kenya has shown that coffee seeds are useful for over two years if stored at 15°C at 41% moisture content in an airtight polythene bag (Mitchell, 45 and Van der Vossen).

Whole bean coffee maintains its freshness the longest. The freshness of ground coffee will be lost in a matter of minutes since its protective cellular structure has been broken and the volatiles have been exposed to the environment to undergo oxidation. An inert gas such as nitrogen can be used to help preserve the whole beans for an extended time. Some people claim that nitrogen could preserve coffee for up to two years, but after a few weeks the coffee no longer acts, tastes, nor smells like freshly roasted coffee.

Whole bean coffee that has been opened and exposed to the environment should be kept for no longer than a week even if a vacuum seal is in place. Coffee that has been roasted very darkly is even more susceptible to oxidation and should be kept for an even shorter period of time. Finally, there is a significant loss in crema development for espresso coffee four days after roasting. Espresso coffee should be purchased immediately after roasting and consumed within a couple of days.

Although not considered part of the processing pipeline proper, nearly all coffee sold to consumers throughout the world is sold as roasted coffee. Consumers can also elect to buy unroasted coffee to be roasted at home.

Coffee Blending is necessary for espresso since a single coffee origin will lack the complexity desired. It is important to remember that the advantage of espresso over other brewing methods is a result of the formation of the crema. Without the crema the espresso would be strong, thick coffee. The crema is an emulsified layer of tiny, smooth bubbles that trap aromatic compounds. This layer coats the tongue and these small bubbles break over time allowing espresso to be enjoyed long after it has been consumed. An aromatic coffee, therefore, is essential to a well prepared espresso blend.

The majority of the espresso blend is made up of a base of coffees from Brazil, the Dominican Republic, Mexico, Peru, Panama or any other origins that contribute a non-overbearing flavour while still contributing to the body and sweetness of the coffee. Small ratios of coffees from Colombia, Costa Rica, Guatemala, and Venezuela are used to add body, acidity, and flavour to a coffee blend. Since these coffees are often very acidic, they are used in small amounts.

To add complexity and brightness when blending espresso, Ethiopian Harrar, Kenyan, Yemen Mocha, Zimbabwe, and Zambian coffees are used. Ethiopian Harrar adds a powerful aroma of blueberries or raspberries while Kenya coffee adds a powerful brightness. To add richness and body when blending for espresso, coffee from the Asian Pacific, such as Sumatra, Sulawesi, Java, East Timor, New Guinea and Ethiopian Yirgacheffe are used. Yirgacheffe has a potent, flowerlike aroma.

Espresso Coffee

— Cup each of the coffees separately. Record aroma, fragrance, flavour, acidity, body, and aftertaste. Cup coffees next to each other to determine which coffees augment the flavour of another. Remember that blending coffee beans is an art and there are no clear espresso blending rules. The goal in espresso blending is that the whole must be greater than the sum of the parts.

— Start with a base of a sweet and heavy bodied Brazilian coffee(s) and add a small amount of another coffee to it. Understand the flavour profile of your base and understand your goal. Ask yourself what coffees might be added to this base to achieve the espresso blend you want. Note the change incurred by adding this coffee and repeat with other coffee origins.

— Next try mixing 3-4 other coffees together until you get a blend that displays the flavour characteristics you desire.
— After determining what type of coffee you would like to use in your espresso blend, begin experimenting with different ratios until you have determined the best ratios to bring out the flavour, sweetness, body and aftertaste desired.
— Experiment with different roasts of each coffee in the blend in the same manner you experimented with adding other coffees to the Brazilian base. Roast one coffee a little lighter or darker than the other coffees in the blend and note any differences. It is usually preferred to roast each coffee separately to its own individual peak and then blend coffee to create the most complexity.

Creating a good espresso blend is not as difficult as it may initially appear. Within a few hours you should have an acceptable espresso blend that will surpass commercial competitors for the simple reason that it is fresher. To perfect the espresso blend by altering the roast to achieve the perfect crema, flavour, acidity, body and aftertaste takes time and patience, but is a rewarding and educational experience. It is often difficult to balance the intense acidity of a shot of espresso which contains a brighter coffee which is used to contribute complexity and liveliness without roasting through the second pop. A successful espresso blend, therefore, should balance this acidity via creative mixing coffee rather than excessive roasting the beans.

Espresso Grinding

To extract espresso properly it is essential to use an espresso burr grinder and to grind per order. The two major types of burrs used are flat and conical. Conical burr coffee grinders are desirable because they increase the surface area of each particle and the amount of flavour that can be extracted from the coffee. Since a conical burr grinder has longer cutting edges, the burrs can rotate at lower speeds, which reduces the heat created. Flat burrs should be replaced after 600 pounds of coffee have been ground and conical burrs should be replaced after 2000 pounds of coffee have been ground.

Coffee Beans

Coffee is freshest immediately after it is ground. After grinding coffee beans, the volatile oils that were previously protected inside the bean are exposed to the air which oxidizes and stales the coffee. This effect occurs immediately after coffee grinding so it is important to tamp and extract the espresso as quickly as possible. The coffee grinder should be activated for 15-20 seconds every time a shot is desired so that only freshly ground coffee is used. Instead of two pulls on the doser, the barista should pull several times until the entire basket is filled with ground coffee.

Espresso coffee should be ground to a size in which the extraction process takes 23-28 seconds. It is important to only adjust the grind and not the pressure one tamps with to control the flow rate. In addition to particle size, the humidity plays a dramatic role on extraction time. Since coffee is hydroscopic, it absorbs moisture from the air causing a tighter pack and longer extraction time. Thus, the grind setting must be changed slightly throughout the day as the barista perceives changes in extraction time.

Espresso Tamping

Espresso Tamping is an art that is often neglected in espresso preparation. The goal is to create a pellet of coffee through which the hot water from the espresso machine will penetrate evenly. Since the water from the espresso machine is under pressure, the espresso pellet must be hard and evenly tamped. The water only knows how to go from a region of high pressure to a region of low pressure. Therefore it is important to prevent paths of least resistance in the coffee pellet and force the water to evenly permeate and extract the coffee.

Correct Espresso Tamping

Once you are done distributing the grounds, it is time for the first tamp. Without moving the porta-filter, hold the espresso tamper so that the base of the handle fits into the palm of your hand. Your wrist should be straight, and the espresso tamper should be a straight extension of your arm. Press gently on the coffee with five pounds of pressure. You will notice that some of the grounds will stick to the side of the basket. Therefore, one must gently tap the basket with the handle on the porta-filter to knock the grounds onto the flat pellet you just formed.

The next step is to apply the finishing espresso tamp. The shape of the pellet has already been formed, and the finishing tamp confirms this impression. With the tamping tool held as before, press on the pellet with thirty pounds of pressure. It is useful to tamp espresso on a bathroom scale until you become comfortable with the amount of force necessary to achieve the appropriate pressure. After tamping, turn the espresso tamper 720° while continuing to apply pressure to polish the surface. Make sure you tamp evenly. An uneven espresso tamp will result in an uneven extraction.

The above steps should be carried out in about thirty seconds. Although speed is important, it is necessary to be careful not to bump the basket during this process. Sharply hitting the basket will unevenly distribute the grounds allowing shortcuts for the water to pass through. As any scientist can appreciate, the path of least resistance is preferred. If there are any weak spots or holes in the espresso pellet the water will push through this area, over extracting this portion of coffee while under extracting the rest of the pellet. Improper espresso tamping will result in a twirling pour or white crema.

Just as correct espresso tamping is essential, so is the use of a proper espresso tamper. The first action you should take is to throw away the plastic round bottom tamper that you currently have. The espresso tamper should be made of aluminum or similar light metal and should have a diameter so it fits firmly into the basket. Marzocco baskets are 58mm, so order the appropriate size. Without a flat packing surface, you create indents which cause uneven extraction.

References

Aguilera J.M AND Stanley D.W., *Microstructural Principles of Food Processing and Engineering,* 2nd Edition, Gaithersburg, MD: Aspen Publishing, 1999.

Archer, G.P., Kennedy, C. J. and Wilson, A. J., "Towards predictive microbiology in frozen food systems-a framework for understanding microbial population dynamics in frozen structures and in freeze-thaw cycles", *International Journal of Food Science and Technology,* 30, pp. 711, 1995.

Blanshard, J. M.V. and Franks, F., *Ice crystallization and its control in frozen-food systems,* Food Structure and behavior, Academic Press, London, UK, pp. 51-65, 1987.

7

Freezing of Fruits and Vegetables

Freezing preservation retains the quality of agricultural products over long storage periods. As a method of long-term preservation for fruits and vegetables, freezing is generally regarded as superior to canning and dehydration, with respect to retention in sensory attributes and nutritive properties. The safety and nutrition quality of frozen products are emphasized when high quality raw materials are used, good manufacturing practices are employed in the preservation process, and the products are kept in accordance with specified temperatures.

Freezing has been successfully employed for the long-term preservation of many foods, providing a significantly extended shelf life. The process involves lowering the product temperature generally to -18 °C or below. The physical state of food material is changed when energy is removed by cooling below freezing temperature. The extreme cold simply retards the growth of microorganisms and slows down the chemical changes that affect quality or cause food to spoil. Competing with new technologies of minimal processing of foods, industrial freezing is the most satisfactory method for preserving quality during long storage periods.

When compared in terms of energy use, cost, and product quality, freezing requires the shortest processing time. Any other conventional method of preservation focused on fruits and vegetables, including dehydration and canning, requires less energy when compared with energy consumption in the freezing process and storage. However, when the overall cost is estimated, freezing costs can be kept as low (or lower) as any other method of food preservation.

Table 1. Frozen food industry in terms of annual sales in 2001

Food items	Sales (million)	US$ % Change vs. 2000
Total Frozen Food Sales	26 600	6.1
Baked Goods	1 400	9.0
Breakfast Foods	1 050	4.1
Novelties	1 900	10.5
Ice Cream	4 500	5.7
Frozen Dessert/Fruit/Toppings	786	5.4
Juices/Drinks	827	-9.7
Vegetables	2 900	4.3

The frozen food market is one of the largest and most dynamic sectors of the food industry. In spite of considerable competition between the frozen food industry and other sectors, extensive quantities of frozen foods are being consumed all over the world. The industry has recently grown to a value of over US$ 75 billion in the U.S. and Europe combined. This number has reached US$ 27.3 billion in 2001 for total retail sales of frozen foods in the U.S. alone. In Europe, based on U.S. currency, frozen food consumption also reached 11.1 million tons in 13 countries in the year 2000. Table 1 represents the division of frozen food industry in terms of annual sales in 2001.

Freezing Technology in Developing Countries

Developed countries, mostly the U.S., dominate the international trade of fruits and vegetables. The U.S. is ranked number one as both importer and exporter, accounting for the highest percent of fresh produce in world trade. However, many developing countries still lead in the export of fresh exotic fruits and vegetables to developed countries. For developing countries, the application of freezing preservation is favourable with several main considerations. From a technical point of view, the freezing process is one of the most convenient and easiest of food preservation methods, compared with other commercial preservation techniques.

The availability of different types of equipment for several different food products results in a flexible process in which degradation of initial food quality is minimal with proper application procedures. As mentioned earlier, the high capital investment of the freezing industry usually plays an important role in terms of economic feasibility of the process in developing countries. As for cost distribution, the freezing process and storage in terms of energy consumption constitute approximately 10 percent of the total cost. Depending on the government regulations, especially in developing countries, energy cost for producers can be subsidized by means of lowering the unit price or reducing the tax percentage in order to enhance production. Therefore, in determining the economical convenience of the process, the cost related to energy consumption (according to energy tariffs) should be considered.

Consumer Demand in Developing Countries

The proportion of fresh food preserved by freezing is highly related to the degree of economic development in a society. As countries become wealthier, their demand for high-valued commodities increases, primarily due to the effect of income on the consumption of high-valued commodities in developing countries. The commodities preserved by freezing are usually the most perishable ones, which also have the highest price. Therefore, the demand for these commodities is less in developing areas. Besides, the need for adequate technology for freezing process is the major drawback of developing countries in competing with industrialised countries.

The frozen food industry requires accompanying developments and facilities for transporting, storing, and marketing their products from the processing plant to the consumer. Thus, a large amount of capital investment is needed for these types of facilities. For developing countries, especially in rural or semi-rural areas, the frozen food industry has therefore not been developed significantly compared to other countries.

In recent years, due to the changing consumer profile, the frozen food industry has changed significantly. The major trend in consumer behaviour documented over the last half century has been the increase in the number of working women and the decline in the family size. These two factors resulted in a reduction in time spent preparing food.

The entry of more women into the workforce also led to improvements in kitchen appliances and increased the variability of ready-to-eat or frozen foods available in the market. Besides, the increased usage of microwave ovens, affecting food habits in general and the frozen food market in particular, as well as allowing rapid preparation of meals and greater flexibility in meal preparation. The frozen food industry is now only limited by imagination, an output of which increases continuously to supply the increasing demand for frozen products and variability.

Today in modern society, frozen fruits and vegetables constitute a large and important food group among other frozen food products. The historical development of commercial freezing systems designed for special food commodities helped shape the frozen food market. Technological innovations as early as 1869 led to the commercial development and marketing of some frozen foods. Early products saw limited distribution through retail establishments due to insufficient supply of mechanical refrigeration. Retail distribution of frozen foods gained importance with the development of commercially frozen vegetables in 1929.

The frozen vegetable industry mostly grew after the development of scientific methods for blanching and processing in the 1940s. Only after the achievement of success in stopping enzymatic degradation, did frozen vegetables gain a strong retail and institutional appeal. Today, market studies indicate that considering overall consumption of frozen foods, frozen vegetables constitute a very significant proportion of world frozen-food categories (excluding ice cream) in Austria, Denmark, Finland, France, Germany, Italy, Netherlands, Norway, Sweden, Switzerland, UK, and the USA. The division of frozen vegetables in terms of annual sales in 2001 is shown in Table 2.

Commercialisation history of frozen fruits is older than frozen vegetables. The commercial freezing of small fruits and berries began in the eastern part of the U.S. in about 1905. The main advantage of freezing preservation of fruits is the extended usage of frozen fruits during off-season. Additionally, frozen fruits can be transported to remote markets that could not be accessed with fresh fruit. Also, freezing preservation makes year-round further processing of fruit products possible, such as jams, juice, and syrups from frozen whole fruit, slices, or pulps.

Future Trends in Freezing Technology

The frozen food industry is highly based in modern science and technology. Starting with the first historical development in freezing preservation of foods, today, a combination of several factors influences the commercialisation and usage of freezing technology.

Table 2. Frozen vegetables in terms of annual sales in 2001

Vegetables	Sales US$ (million)	% Change vs. 2000
Broccoli	184	4.4
Com/Corn on the Cob	312	3.5
Green Beans	115	6.0
Mixed Vegetables	450	7.2
Peas	207	3.9
Potatoes	1070	4.4

The future growth of frozen foods will mostly be affected by economical and technological factors. Growth in population, personal incomes, relative cost of other forms of foods, changes in tastes and preferences, and technological advances in freezing methods are some of the factors concerned with the future of freezing technology.

Population growth and increasing demand for food has generated the need for commercial production of food commodities in large-scale operations. Thus, availability of proper equipment suitable for continuous processing would be valuable for freezing preservation methods. In addition depending on personal incomes, relative cost of frozen products is one of the most important of economical factors. Producing the highest quality at the lowest cost possible is highly dependent on the technology used. As a result, developments in freezing technology in recent years have mostly been characterised by the improvements in mechanical handling and process control to increase freezing rate and reduce cost.

Today an increasing demand for frozen foods already exits and further expansion of the industry is primarily dependent on the ability of food processors to develop higher qualities in both process techniques and products. Improvements can only be achieved by focusing on new technologies and investigating poorly understood factors that influence the quality of frozen food products. Improvements in new and convenient forms of foods, as well as more information on relative cost and nutritive values of frozen foods, will contribute toward continued growth of the industry.

Method of Preservation

Freezing is a widely used method of food preservation based on several advantages in terms of retention of food quality and ease of process. Beginning with the earliest history of freezing, the technology has been highly affected over the years by the developments and improvements in freezing techniques. In order to understand and handle the concepts associated with freezing of foods, it is necessary to examine the fundamental factors governing the freezing process.

Freezing has long been used as a method of preservation, and history reveals it was mostly shaped by the technological developments in the process. A small quantity of ice produced without using a "natural cold" in 1755 was regarded as the first milestone in the freezing process. Firstly, ice-salt systems were used to preserve fish and later on, by the late 1800's, freezing was introduced into large-scale operations as a method of commercial preservation. Meat, fish, and butter, the main products preserved in this early example, were frozen in storage chambers and handled as bulk commodities.

In the following years, scientists and researchers continuously worked to achieve success with commercial freezing trials on several food commodities. Among these commodities, fruits were one of the most important since freezing during the peak growing season had the advantage of preserving fruit for later processing into jams, jellies, ice cream, pies, and other bakery foods. Although commercial freezing of small fruits and berries first began around 1905 in the eastern part of the United States, the commercial freezing of vegetables is much more recent. Starting from 1917, only private firms conducted trials on freezing vegetables, but achieving good quality in frozen vegetables was not possible without pre-treatments due to the enzymatic deterioration. In 1929, the necessity of blanching to inactivate enzymes before freezing was concluded by several researchers to avoid deterioration and off-flavours caused by enzymatic degradation.

The modern freezing industry began in 1928 with the development of double-belt contact freezers by a technologist named Clarence Birdseye. After the revolution in the quick freezing process and equipment, the industry became more flexible, especially with the usage of multi-plate freezers. The earlier methods achieved successful freezing of fish and poultry, however with the new quick freezing system, packaged foods could be frozen between two metal belts as they moved through a freezing tunnel. This improvement was a great advantage in the commercial large-scale freezing of fruits and vegetables. Furthermore, quick-freezing of consumer-size packages helped frozen vegetables to be accepted rapidly in late 1930s.

Today, freezing is the only large-scale method that bridges the seasons, as well as variations in supply and demand of raw materials such as meat, fish, butter, fruits, and vegetables. Besides, it makes possible movement of large quantities of food over geographical distances. It is important to control the freezing process, including the pre-freezing preparation and post-freezing storage of the product, in order to achieve high-quality products. Therefore, the theory of the freezing process and the parameters involved should be understood clearly.

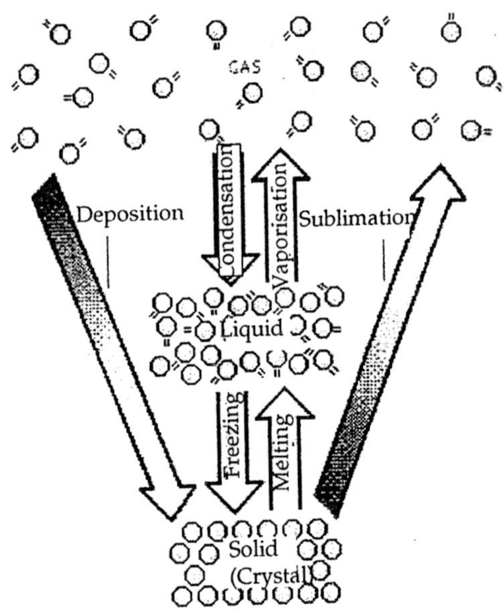

Figure 1. A schematic illustration of overall freezing process.

The freezing process mainly consists of thermodynamic and kinetic factors, which can dominate each other at a particular stage in the freezing process. Major thermal events are accompanied by reduction in heat content of the material during the freezing process as is shown in Figure 1.

The material to be frozen first cools down to the temperature at which nucleation starts. Before ice can form, a nucleus, or a seed, is required upon which the crystal can grow; the process of producing this seed is defined as nucleation. Once the first crystal appears in the solution, a phase change occurs from liquid to solid with further crystal growth. Therefore, nucleation serves as the initial process of freezing, and can be considered as the critical step that results in a complete phase change.

Freezing Point of Foods

Freezing point is defined as the temperature at which the first ice crystal appears and the liquid at that temperature is in equilibrium with the solid. If the freezing point of pure water is considered, this temperature will correspond to 0 °C (273°K). However, when food systems are frozen, the process becomes more complex due to the existence of both free and bound water. Bound water does not freeze even at very low temperatures. Unfreezable water contains soluble solids, which cause a decrease in the freezing point of water lower than 0 °C. During the freezing process, the concentration of soluble solids increases in the unfrozen water, resulting in a variation in freezing temperature. Therefore, the temperature at which the first ice crystal appears is commonly regarded as the initial freezing temperature. There are empirical equations in literature that can calculate the initial freezing temperature of certain foods as a function of their moisture content.

There are several methods of food freezing, and depending on the method used, the quality of the frozen food may vary. However, regardless of the method chosen, the main principle behind all freezing processes is the same in terms of process parameters. The International Institute of Refrigeration (IIR) has provided definitions to establish a basis for the freezing process. According to their definition, the freezing process is basically divided into three stages based on major temperature changes in a particular location in the product, as shown in Figures 2 and 3 for pure water and food respectively.

Beginning with the prefreezing stage, the food is subjected to the freezing process until the appearance of the first crystal. If the material frozen is pure water, the freezing temperature will be 0 °C and, up to this temperature, there will be a subcooling until the ice formation begins. In the case of foods during this stage, the temperature decreases to below freezing temperature and, with the formation of the first ice crystal, increases to freezing temperature.

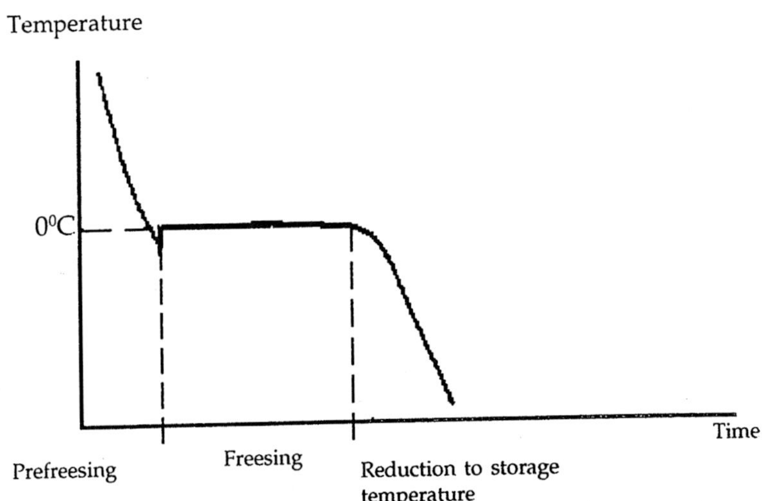

Figure 2. Practical definition of the freezing process for pure water.

The second stage is the freezing period; a phase change occurs, transforming water into ice. For pure water, temperature at this stage is constant; however, it decreases slightly in foods, due to the increasing concentration of solutes in the unfrozen water portion. The last stage starts when the product temperature reaches the point where most freezable water has been converted to ice, and ends when the temperature is reduced to storage temperature.

The freezing time and freezing rate are the most important parameters in designing freezing systems. The quality of the frozen product is mostly affected by the rate of freezing, while time of freezing is calculated according to the rate of freezing.

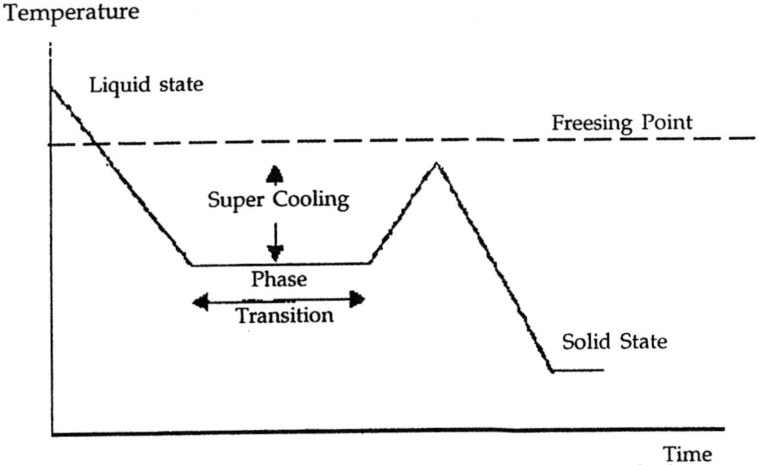

Figure 3. Practical definition of the freezing process for foods.

For industrial applications, they are the most essential parameters in the process when comparing different types of freezing systems and equipment.

Freezing Time

Again, freezing time is one of the most important parameters in the freezing process, defined as time required to lower product temperature from its initial temperature to a given temperature at its thermal center. Since the temperature distribution within the product varies during freezing process, the thermal center is generally taken as reference.

Thus, when the geometrical center of the product reaches the given final temperature, this ensures the average product temperature has been reduced to a storage value. Freezing time depends on several factors, including the initial and final temperatures of the product and the quantity of heat removed, as well as dimensions (especially thickness) and shape of product, heat transfer process, and temperature. The International Institute of Refrigeration defines various factors of freezing time in relation to both the product frozen and freezing equipment. The most important are:

- Dimensions and shape of product, particularly thickness
- Initial and final temperatures
- Temperature of refrigerating medium
- Surface heat transfer coefficient of product
- Change in enthalpy
- Thermal conductivity of product

Calculation of freezing time in food systems is difficult in comparison to pure systems since the freezing temperature changes continuously during the process. Using a simplified approach, time elapsed between initial freezing until when the entire product is frozen can be regarded as the freezing time. Plank's equation is commonly used to estimate freezing time, however due to assumptions involved in the calculation it is only useful for obtaining an approximation of freezing time. The derivation of the equation starts with the assumption the product being frozen is initially at freezing temperature. Therefore, the calculated freezing time represents only the freezing period. The equation can be further modified for different geometries including slab, cylinder, and sphere, where for each geometry, the coefficients are arranged in relation to the dimensions.

Table 3. Coefficients P and R of Equation 1

Geometry	P	R	Dimension
Infinite slab	1/2	1/8	thicknesse
Infinite cylinder	1/4	1/16	radiusr
Sphere	1/6	1/24	radiusr

where l_1 is the latent heat of frozen fraction, k and r are the thermal conductivity and density of the frozen layer, while h is the coefficient of heat transfer by convection to the exterior. T_f denotes the body temperature of the product when introduced into a freezer in which the external temperature is T_e The coefficients R and P are given in Table 3 and arranged according to the geometry of the product frozen, where the letter e denotes the dimension.

$$t_F = \frac{\rho \lambda_1}{T_F - T_e}\left[\frac{e^2 R}{k} + \frac{eP}{h}\right] \tag{1}$$

As mentioned earlier, the equation of Plank assumes the food is at a freezing temperature at the beginning of the freezing process. However, the food is usually at a temperature higher than freezing temperature. The real freezing time should therefore be the sum of time calculated from the equation of Plank and the time needed for the product's surface to decrease from initial temperature to freezing temperature.

Several works have attempted to calculate real freezing time, as in one presented by Nagaoka et al. Nagaoka's equation (Eq. 2) calculates the amount of heat elimination required to decrease a product's temperature from initial temperature to freezing temperature, as well as the amount of heat released during the phase change and the amount of heat eliminated to reach freezing temperature. Further empirical equations can be found in literature in detail.

$$t_F = \frac{\rho \Delta H}{T_F - T_e}\left[\frac{Re^2}{k} + \frac{Pl}{h}\right][1 + 0.008(T_i - T_F)] \tag{2}$$

where T_i is the temperature of the food at the initiation of freezing, DH is the difference between the enthalpy of the food at initial temperature and end of freezing. Re and Pl are the dimensionless numbers, while k and h are the thermal conductivity and the coefficient of heat transfer, respectively.

For calculating freezing time of products with irregular shape, a common property of most food products—especially fruits and vegetables—a dimensionless factor has been employed in equations.

Freezing Rate

The freezing rate (°C/h) for a product or package is defined as the ratio of difference between initial and final temperature of product to freezing time. At a particular location within the product, a local freezing rate can be defined as the ratio of the difference between the initial temperature and desired temperature to the time elapsed in reaching the given final temperature. The quality of frozen products is largely dependent on the rate of freezing.

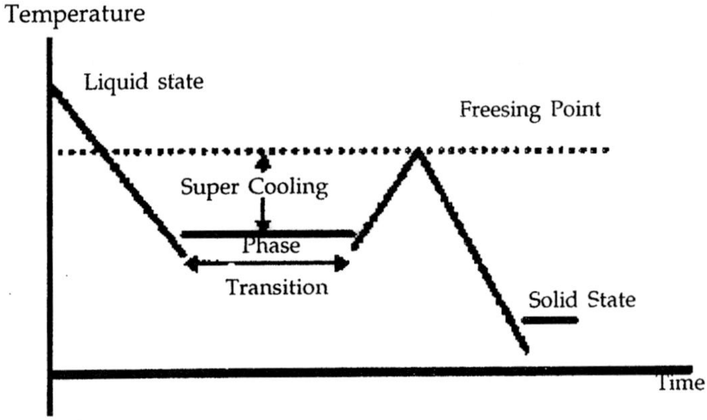

Figure 4. Freezing preservation dynamics curve.

Rapid freezing results in better quality frozen products when compared with slow freezing. If freezing is instantaneous, there will be more locations within the food where crystallisation begins. In contrast, if freezing is slow, the crystal growth will be slower with few nucleation sites resulting in larger ice crystals. Large ice crystals are known to cause mechanical damage to cell walls in addition to cell dehydration. Thus, the rate of freezing for plant tissues is extremely important due to the effect of freezing rate on the size of ice crystals, cell hydration, and damage to cell walls. The figure 4 shows a general behaviour of the dynamics curve of freezing preservation.

Rapid freezing is advantageous for freezing of many foods, however some products are susceptible to cracking when exposed to extremely low temperature for long periods. Several mechanisms, including volume expansion, contraction and expansion, and building of internal pressure, are proposed in literature explaining the mechanisms of product damage during freezing.

Energy Requirements

For fruits and vegetables, the amount of energy required for freezing is calculated based on the enthalpy change and the amount of product to be frozen. The following equation is reported by Riedel for calculation of refrigeration requirements for fruits and vegetables.

$$\Delta H = [1 - \frac{X_{SNJ}}{100}]\Delta H_J + 1.21[\frac{X_{SNJ}}{100}]\Delta T \quad (3)$$

X_{SNJ}: Percentage of the product solids different from juice (Dry matter fraction of the juice)

ΔH_j: Enthalpy change during freezing of the juice fraction

ΔT: Temperature difference between initial and final temperature of the product

Define Refrigeration

Refrigeration is defined as the elimination of heat from a material at a temperature higher than the temperature of its surroundings. The mechanism of refrigeration is a part of the freezing process and freezing storage involved in the thermodynamic aspects of freezing. According to the second law of thermodynamics, heat only flows from higher to lower temperatures. Therefore, in order to raise the heat from a lower to a higher temperature level, expenditure of work is needed. The aim of industrial refrigeration processes is to eliminate heat from low temperature points towards points with higher temperature. For this reason, either closed mechanical refrigeration cycles in which refrigeration fluids circulate, or open cryogenic systems with liquid nitrogen (LIN) or carbon dioxide (CO_2), are commonly used by the food industry.

The main elements in a closed mechanical refrigeration system are the condenser, compressor, evaporator, and the expansion valve. The refrigerants hydrochlorofluorocarbon

(HCFC) and ammonia are examples of the refrigerants circulated in these types of mechanical refrigeration systems. A simple scheme for the closed mechanical refrigeration system is shown in Figure 5.

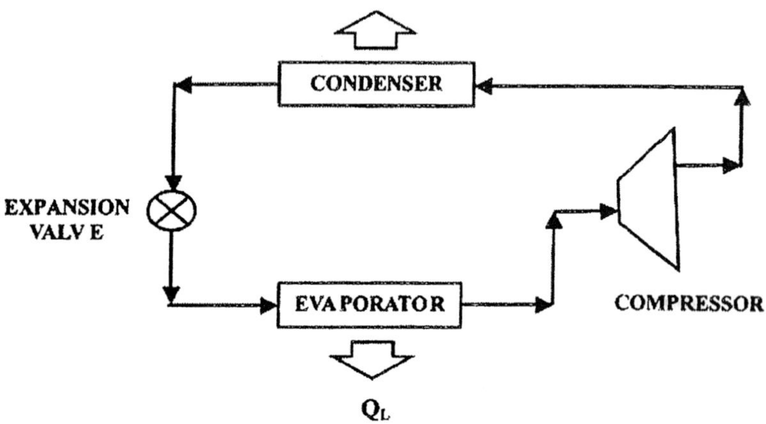

Figure 5. A simple scheme for a one-stage closed mechanical refrigeration system.

Starting at the suction point of the compressor, fluid in a vapour state is compressed into the compressor where an increase in temperature and pressure takes place. The fluid then flows through the condenser where it decreases in energy by giving off heat and converting to a liquid state. After the phase, a change occurs inside the condenser, the fluid flows through the expansion valve where the pressure decreases to convert liquid into a form of liquid-gas mixture. Finally, the liquid-gas mixture flows through the evaporator where it is converted into a saturated vapour state and removes heat from the environment in the process of cooling. With this last stage the loop restarts again. The other refrigeration system employed by the food industry is the cryogenic system with carbon dioxide or liquid nitrogen. The refrigerant in this system is consumed differently from the circulating fluid in closed mechanical systems.

There are several refrigerants available for refrigeration systems. The selection of a proper refrigerant is based on physical, thermodynamic, and chemical properties of the fluid. Environmental considerations are also important in refrigerant selection, since leaks within the system produce deleterious effects on the atmospheric ozone layer. Some refrigerants, including halocarbons, have been banned to avoid potential hazardous effects. For industrial applications, ammonia is commonly used, while chlorofluoromethane and tetrafluoroethane are also recommended as refrigerants.

Freezing Capacity

Freezing equipment selection is based on the requirements for freezing a certain quantity of food per hour. For any type of freezer, freezing capacity (expressed in tonnes per hour) is defined as the ratio of the quantity of the product that can be loaded into the freezer to the holding time of the product in that particular freezer. The first parameter, the amount of food product loaded into the freezer, is affected by both the dimensions of the product and the mechanical constraints of the freezer. The denominator (holding time) has an important role in freezing systems and is based on the calculation of the amount of heat removed from the product per hour, which varies depending on the type of product frozen.

Types of Freezing Systems

There is a variety of freezing systems available for freezing, and for most products, more than one type of freezer can be used. Therefore, in selecting a freezing system initially, a cost-benefit analysis should be conducted based on three important factors: economics, functionality, and feasibility. Financial considerations mainly involve capital investment and the production cost of selected equipment. Product losses during freezing operation should be included in cost estimation since generating higher cost freezers may have other benefits in terms of reducing product losses.

Functional factors are mostly based on the suitability of the selected freezer for particular products. The mode of process, either in-line or batch, should be considered based on the fact that computerised systems are becoming more important for ease of handling and lowering production costs. Mechanical constraints for the freezer should also be considered since some types of freezers are not physically suitable for freezing certain products. Lastly, the feasibility of the process should be considered in terms of plant location or location of the processing area, as well as cleanability and hygienic design, and desired product quality.

These factors and initial considerations can help eliminate several choices in freezer selection, but the relative importance of factors may change depending on the process. For developing countries where the freezing application is relatively new, the cost factor becomes more important than other factors due to the decreased production rates and need for lower capital investment costs.

Use of Freezing Equipment

The industrial equipment for freezing can be categorised in many ways, namely as equipment used for batch or in-line operation, heat transfer systems (air, contact, cryogenic), and product stability. The rate of heat transfer from the freezing medium to the product is important in defining the freezing time of the product. Therefore, the equipment selected for freezing process characterises the rate of freezing.

Air Blast Freezer

The air blast freezer is one the oldest and commonly used freezing equipment due to its temperature stability and versatility for several product types. In general, air is used as the freezing medium in the freezing design, either as still air or forced air. Freezing is accomplished by placing the food in freezing rooms called sharp freezers. Still, air freezing is the cheapest way of freezing and has the added advantage of a constant temperature during frozen storage, which allows usage for unprocessed bulk products like beef quarters and fish.

However, it is the slowest method of freezing due to the low surface heat transfer coefficient of circulating air inside the room. Freezing time in sharp freezers is largely dependent on the temperature of the freezing chamber and the type, initial temperature, and size of product. An improved version of the still air freezer is the forced air freezer, which consists of air circulation by convection inside the freezing room. However, even modification of the sharp freezer with extra refrigeration capacity and fans for increased air circulation does not help control the air flow over the products during slow freezing.

There are a considerable number of designs and arrangements for air blast freezers, primarily grouped in two categories depending on the mode of process, as either inline or batch. Continuous freezers are the most suitable systems for mass production of packaged products with similar freezing times, in which the product is carried through on trucks or on conveyors. The system works on a semi-batch principle when trucks are used, since they remain stationary during the process except when a new truck enters one end of the tunnel, thus moving the others along to release a finished one at the exit. The batch freezers are more flexible since a variety of products can be frozen at the same time on individual trolleys. Over-loading may be a problem for these types of freezers, thus the process requires closer supervision than continuous systems.

Tunnel Freezers

In tunnel freezers, the products on trays are placed in racks or trolleys and frozen with cold air circulation inside the tunnel. In order to allow air circulation, optimum space is provided between layers of trolley, which can be moved continuously in and out of the freezer manually or by forklift trucks. This freezing system is suitable for all types of products, although there are some mechanical constraints including the requirement of high manpower for handling, cleaning, and transportation of trays.

Belt Freezers

Belt freezers were first designed to provide continuous product flow with the help of a wire mesh conveyor inside the blast rooms. A poor heat transfer mechanism and the mechanical problems were solved in modern belt freezers by providing a vertical airflow to force air through the product layer. Airflow has good contact with the product only when the entire product is evenly distributed over the conveyor belt.

In order to decrease required floor space, the belts can be arranged in a multi-tier belt freezer or a spiral belt freezer. Spiral belt freezers consist of a belt that can be bent laterally around a rotating drum to maximize belt surface area in a given floor space.

Figure 6. The cross-section view of a spiral belt freezer.

This type of design has the advantage of eliminating product damage in transfer points, especially for products that require gentle handling. Both packed and unpacked products with long freezing times (10 min to 3 hr) can be frozen in spiral belt freezers due to the flexibility of the equipment. A typical spiral belt freezer is shown in Figure 6.

Fluidized Bed Freezers

The fluidized bed freezer, a fairly recent modified type of air-blast freezer for particular product types, consists of a bed with a perforated bottom through which cold air is blown vertically upwards. The system relies on forced cold air from beneath the conveyor belt, causing the products to suspend or float in the cold air stream. The use of high air velocity is very effective for freezing unpacked foods, especially when they can be completely surrounded by flowing air, as in the case of fluidized bed freezers.

The use of fluidisation has several advantages compared with other methods of freezing since the product is individually quick frozen (IQF), which is convenient for particles with a tendency to stick together. The idea of individually quick frozen foods (IQF) started with the first technological developments aimed at quick freezing. The need for an effective means of freezing small particles with the potential for lumping during the process is the objective of IQF freezing. Small vegetables, prawns, shrimp, french-fried potatoes, diced meat, and fruits are some of the products now frozen with this technology.

Contact Freezers

Contact freezing is the one of the most efficient ways of freezing in terms of heat transfer mechanism.

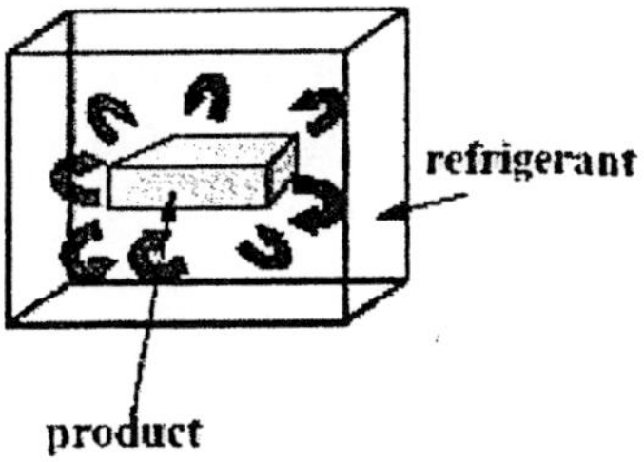

Figure 7. Direct contact freezer.

In the process of freezing, the product can be in direct or indirect contact with the freezing medium. For direct contact freezers, the product being frozen is fully surrounded by the freezing medium, the refrigerant, maximizing the heat transfer efficiency. A schematic illustration is given in Figure 7. For indirect contact freezers, the product is indirectly exposed to the freezing medium while in contact with the belt or plate, which is in contact with the freezing medium.

Immersion Freezers

The immersion freezer consists of a tank with a cooled freezing media, such as glycol, glycerol, sodium chloride, calcium chloride, and mixtures of salt and sugar. The product is immersed in this solution or sprayed while being conveyed through the freezer, resulting in fast temperature reduction through direct heat exchange. Direct immersion of a product into a liquid refrigerant is the most rapid way of freezing since liquids have better heat conducting properties than air.

The solute used in the freezing system should be safe without taste, odour, colour, or flavour, and for successful freezing, products should be greater in density than the solution. Immersion freezing systems have been commonly used for shell freezing of large particles due to the reducing ability of product dehydration when the outer layer is frozen quickly. A commonly seen problem in these freezing systems is the dilution of solution with the product, which can change the concentration and process parameters. Thus, in order to avoid product contact with the liquid refrigerant, flexible membranes can be used. A simple illustration of the immersion freezer is shown in Figure 8.

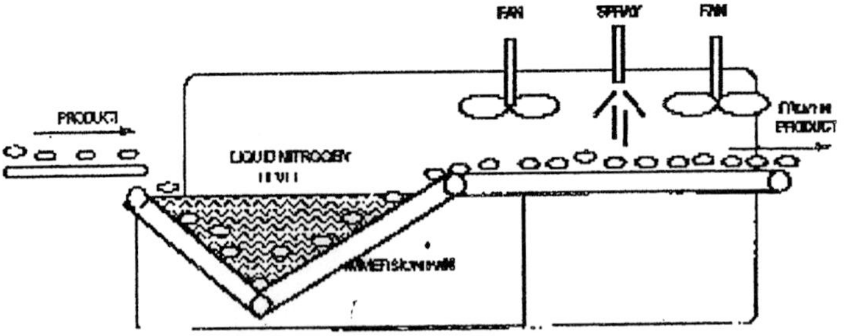

Figure 8. Simple illustration of a typical immersion freezer.

Indirect Contact Freezers

In this type of freezer, materials being frozen are separated from the refrigerant by a conducting material, usually a steel plate. The mechanism of indirect contact freezer is shown in Figure 9. Indirect contact freezers generally provide an efficient medium for heat transfer, although the system has some limitations, especially when used for packaged foods due to resistance of package to heat transfer.

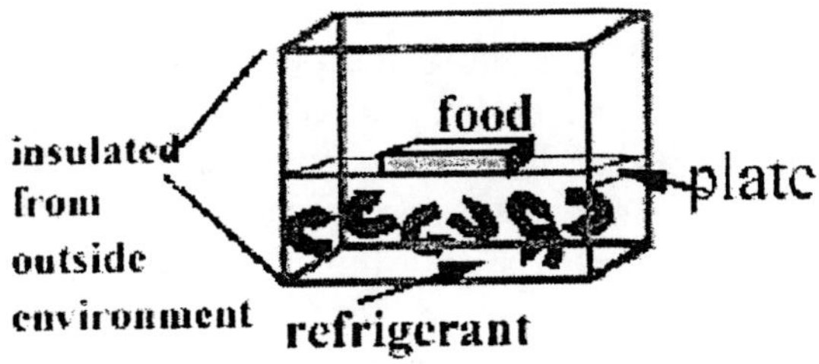

Figure 9. Indirect contact freezer.

Plate Freezers

The most common type of contact freezer is the plate freezer. In this case, the product is pressed between hallow metal plates, either horizontally or vertically, with a refrigerant circulating inside the plates. This type of freezing system is only limited to regular-shaped materials or blocks like beef patties or block-shaped packaged products.

Contact Belt Freezers

This type of freezer is designed with single-band or double-band for freezing of thin product layers. The design can be either straight forward or drum. Typical products frozen in belt freezers are, fruit pulps, egg yolk, sauces and soups.

Cryogenic Freezers

Cryogenic freezing is a relatively new method of freezing in which the food is exposed to an atmosphere below -60 °C through direct contact with liquefied gases such as nitrogen or carbon dioxide. This type of system differs from other freezing systems since it is not connected to a refrigeration plant; the refrigerants used are liquefied in large industrial installations and shipped to the food-freezing factory in pressure vessels. Thus, the small size and mobility of cryogenic freezers allow for flexibility in design and efficiency of the freezing application. Low initial investment and rather high operating costs are typical for cryogenic freezers.

Liquid Nitrogen Freezers

Liquid nitrogen, with a boiling temperature of -196 °C at atmospheric pressure, is a by-product of oxygen manufacture. The refrigerant is sprayed into the freezer and evaporates both on leaving the spray nozzles and on contact with the products. The system is designed in a way that the refrigerant passes in counter current to the movement of the products on the belt giving high transfer efficiency. The refrigerant consumption is in the range of 1.2-kg refrigerant per kg of the product. Typical food products used in this system are, fish fillets, seafood, fruits, berries.

Liquid Carbon Dioxide Freezers

Liquid carbon dioxide exists as either a solid or gas when stored at atmospheric pressure. When the gas is released to the atmosphere at -70 °C, half of the gas becomes dry-ice snow and the other half stays in the form of vapour. This unusual property of liquid carbon dioxide is used in a variety of freezing systems, one of which is a pre-freezing treatment before the product is exposed to nitrogen spray.

Packaging of Frozen Food

Proper packaging of frozen food is important to protect the product from contamination and damage while in transit from the manufacturer to the consumer, as well as to preserve food value, flavour, colour, and texture. There are several factors considered in designing a suitable package for a frozen food. The package should be attractive to the consumer, protected from external contamination, and effective in terms of processing, handling, and cost. Proper selection is based on the type of package and material. There are typically three types of packaging used for frozen foods: primary, secondary, and tertiary. The primary package is in direct contact with the food and the food is kept inside the package up to the time of use. Secondary packaging is a form of multiple packaging used to handle packages together for sale. Tertiary packaging is used for bulk transportation of products.

Packaging materials should be moisture-vapour-proof to prevent evaporation, thus retaining the highest quality in frozen foods. Oxygen should also be completely evacuated from the package using a vacuum or gas-flush system to prevent migration of moisture and oxygen. Glass and rigid plastic are examples of moisture-vapour-proof packaging materials. Many packaging materials, however, are not moisture-vapour-proof, but are sufficiently moisture-vapour-resistant to retain satisfactory quality in foods. Most bags, wrapping materials, and waxed cartons used in freezing packaging are moisture-vapour-resistant. In general, the containers should be leakage free while easy to seal. Durability of the material is another important factor to consider, since the packaging material must not become brittle at low temperatures and crack.

A range of different packaging materials, mainly grouped as rigid and non-rigid containers, can be used for primary packaging. Glass, plastic, tin, and heavily waxed cardboard materials are in the rigid container group and usually used for packaging of liquid food products. Glass containers are mostly used for fruits and vegetables if they are not water-packed. Plastics are the derivatives of the oil-cracking industry. Non-rigid containers include bags and sheets made of moisture-vapour-resistant heavy aluminum foil, polyethylene or laminated papers. Bags are the most commonly used packaging materials for frozen fruits and vegetables due to their flexibility during processing and handling. They can be used with or without outer cardboard cartons to protect against tearing.

Shape and size of the container are also important factors in freezing products. Serving size may vary depending on the type of product and selection should be based on the amount of food determined for one meal. For shape of the container, freezer space must be considered since rigid containers with flat tops and bottoms stack well in the freezer, while round containers waste freezer space.

Storage and Distribution

The quality of the final product depends on the history of the raw material. Using the lowest possible temperature is essential for frozen storage, transport, and distribution in achieving a high-quality product, since deteriorative processes are mainly temperature dependent. The lower the product temperature is, the slower the speed of reaction is leading to loss of quality. The temperatures of supply chains in freezing applications from the factory to the retail cabinet should be carefully monitored. The temperature regime covering the freezing process, the cold-store temperatures (£ -18 °C), distribution temperatures (£ -15 °C), and retail display (£ -12 °C) are given as legal standards.

Freezing Fruits and Vegetables in Small and Medium Scale Operations

The preservation of fruits and vegetables by freezing is one of the most important methods for retaining high quality in agricultural products over long-term storage. In particular, the freshness qualities of raw fruits and vegetables can be retained for long periods, extending well beyond the normal season of most horticultural crops. The potential application of freezing preservation of fruits and vegetables, including tropical products, has been increasing recently in parallel with developments in developing countries.

Freezing Fruits

The effect of freezing, frozen storage, and thawing on fruit quality has been investigated over several decades. Today frozen fruits constitute a large and important food group. The quality demanded in frozen fruit products is mostly based on the intended use of the product. If the fruit is to be eaten without any further processing after thawing, texture characteristics are more important when compared to use as a raw material in other industries. In general, conventional methods of freezing tend to destroy the turgidity of living cells in fruit tissue.

Different from vegetables, fruits do not have a fibrous structure that can resist this destructive effect. Additionally, fruits to be frozen are harvested in a fully ripe state and are soft in texture. On the contrary, a great number of vegetables are frozen in an immature state. Fruits have delicate flavours that are easily damaged or changed by heat, indicating they are best eaten when raw and decrease in quality with processing. In the same way, attractive colour is important for frozen fruits. Chemical treatments or additives are often used to inactivate the deteriorative enzymes in fruits. Therefore, proper processing is essential for all steps involved, from harvesting to packaging and distribution.

Production

The characteristics of raw materials are of primary importance in determining the quality of the frozen product. These characteristics include several factors such as genetic make-up, climate of the growing area, type of fertilization, and maturity of harvest. The ability to withstand rough handling, resistance to virus diseases, molds, uniformity in ripening, and yield are some of the important characteristics of fruits in terms of economical aspects considered in production. The use of mechanical harvesting generally causes bruising of fruits and results in a wide range of maturity levels for fruits. In contrast, hand-picking provides gentler handling and maturity sorting of fruits. However in most cases, it is non-economical compared to mechanical harvesting due to high labour cost.

As a rule, harvesting of fruits at an optimum level for commercial use is difficult. Simple tests like pressure tests are applied to determine when a fruit has reached optimum maturity for harvest. Colour is also one of the characteristics used in determining maturity since increased maturation causes a darker colour in fruits. A combination of colour and pressure tests is a better way to assess maturity level for harvesting. Controlled atmosphere storage is a common method of storage for some fruits prior to freezing.

In principle, a controlled atmosphere high in carbon dioxide and low in oxygen content slows down the rate of respiration, which may extend shelf life of any respiring fruit during storage. Due to the fact that these fruits do not ripen appreciably after picking, most fruits are picked as near to eating-ripe maturity as possible.

Pre-process Handling

Freezing preservation of fruits can only help retain the inherent quality present initially in a product since the process does not improve the quality characteristics of raw materials. Therefore, quality level of the raw materials prior to freezing is the major consideration for successful freezing. Washing and cutting generally results in losses when applied after thawing. Thus, fruits should be prepared prior to the freezing process in terms of peeling, slicing or cutting. Freezing preservation does not require specific unit operations for cleaning, rinsing, sorting, peeling, and cutting of fruits.

Fruits that require peeling before consumption should be peeled prior to freezing. Peeling is done by scalding the fruit in hot water, steam or hot lye solutions. The effect of peeling on the quality of frozen products has been studied for several fruits, including kiwi, banana, and mango. The rate of freezing can be increased by decreasing the size of products frozen, especially for large fruits. An increase in the freezing rate results in smaller ice crystals, which decreases cellular damage in fruit tissue. Banana, tomato, mango, and kiwi are some examples of large fruits commonly cut into smaller cubes or slices prior to freezing.

The objective of blanching is to inactivate the enzymes causing detrimental changes in colour, odour, flavour, and nutritive value, but heat treatment causes loss of such characteristics in fruits. Therefore, only a few types of fruits are blanched for inactivation of enzymes prior to freezing. The loss of water-soluble minerals and vitamins during blanching should also be minimized by keeping blanching time and temperature at an optimum combination.

Addition of sugars is an extremely important pretreatment for fruits prior to freezing since the treatment has the effect of excluding oxygen from the fruit, which helps to retain colour and appearance. Sugars when dissolved in solutions act by withdrawing water from cells by osmosis, resulting in very concentrated solutions inside the cells. The high concentration of solutes depresses the freezing point and therefore reduces the freezing within the cells, which inhibits excessive structural damage. Sugar syrups in the range of 30-60 percent sugar content are commonly used to cover the fruit completely, acting as a barrier to oxygen transmission and browning. Several experiments have shown the protective effect of sugar on flavour, odour, colour, and nutritive value during freezing, especially for frozen berries.

Fruits exposed to oxygen are susceptible to oxidative degradation, resulting in browning and reduced storage life of products. Therefore, packaging of frozen fruits is based on excluding air from the fruit tissue. Replacement of oxygen with sugar solution or inert gas, consuming the oxygen by glucose-oxidase and/or the use of vacuum and oxygen-impermeable films are some of the methods currently employed for packaging frozen fruits. Plastic bags, plastic pots, paper bags, and cans are some of the most commonly used packaging materials (with or without oxygen removal) selected, based on penetration properties and thickness.

There are several types of fruit packs suitable for freezing: syrup pack, sugar pack, unsweetened pack, and tray pack and sugar replacement pack. The type of pack is usually selected according to the intended use for the fruit. Syrup-packed fruits are generally used for cooking purposes, while dry-packed and tray-packed fruits are good for serving raw in salads and garnishes.

The proportion of sugar to water used in a syrup pack depends on the sweetness of the fruit and the taste preference of the consumer. For most fruits, 40 percent sugar syrup is recommended. Lighter syrups are lower in calories and mostly desirable for miid-flavoured fruits to prevent masking the flavour, while heavier syrups may be used for very sour fruits.

Syrup is prepared by dissolving the sugar in warm water and cooling the solution down before usage. Just enough cooled syrup is used to cover the prepared fruit after it has been settled by jarring the container. In order to keep the fruit under the syrup, a small piece of crumpled waxed paper or other water resistant wrapping material is placed on top; the fruit is pressed down into the syrup before closing, then sealed and frozen.

Pectin can be used to reduce sugar content in syrups when freezing berries, cherries, and peaches. Pectin syrups are prepared by dissolving 1 box of powdered pectin with 1 cup of water. The solution is stirred and boiled for 1 minute; 1/2 cup of sugar is added and dissolved; the solution is then cooled down with the addition of cold water. Previously prepared fruit is put into a 4 to 6 quart bowl and enough pectin syrup is added to cover the fruit with a thin film. The pack is sealed and promptly frozen.

In preparing a sugar pack, sugar is first sprinkled over the fruit. Then the container is agitated gently until the juice is drawn out and the sugar is dissolved. This type of pack is generally used for soft sliced fruits such as peaches, strawberries, plums, and cherries, by using sufficient syrup to cover the fruit. Some whole fruits may also be coated with sugar prior to freezing.

Unsweetened packs can be prepared in several ways, either dry-packed, covered with water containing ascorbic acid, or packed in unsweetened juice. When water or juice is used in syrup and sugar packs, fruit is submerged by using a small piece of crumpled water-resistant material. Generally, unsweetened packs yield a lower quality product when compared with sugar packs, with the exception, some fruits such as raspberries, blueberries, scalded apples, gooseberries, currants, and cranberries maintain good quality without sugar.

Unsweetened packs are generally prepared by using tray packs in which a single layer of prepared fruit is spread on shallow trays, frozen, and packaged in freezer bags promptly. The fruit sections remain loose without clumping together, which offers the advantage of using frozen fruit piece by piece.

Artificial sweeteners can be used instead of sugar in the form of sugar substitutes. The sweet taste of sugar can be replaced by using these kinds of sweeteners, however the beneficial effects of sugar like colour protection and thick syrup can not be replaced. Fruits frozen with sugar substitutes will freeze harder and thaw more slowly than fruits preserved with sugar.

Freezing Vegetables

Freezing is often considered the simplest and most natural way of preservation for vegetables. Frozen vegetables and potatoes form a significant proportion of the market in terms of frozen food consumption. The quality of frozen vegetables depends on the quality of fresh products, since freezing does not improve product quality. Pre-process handling, from the time vegetables are picked until ready to eat, is one of the major concerns in quality retention.

The choice of the right cultivar and maturity before crop is harvested are the two most important factors affecting raw material quality. Raw material characteristics are usually related to the vegetable cultivar, crop production, crop maturity, harvesting practices, crop storage, transport, and factory reception. The choice of crop cultivars is mostly based on their suitability for frozen preservation in terms of factory yield and product quality. Some of the characteristics used as selection criteria are as follows:

- Suitability for mechanical harvesting
- Uniform maturity

— Exceptional flavour and uniform colour and desirable texture

— Resistance to diseases

— High yield

Although cultivar selection is a major factor affecting the quality of the final product, many practices in the field and factors during growth of crop can also have a significant effect on quality. Those practices include site selection for growth, nutrition of crop, and use of agricultural chemicals to control pests or diseases. The maturity assessment for harvesting is one of the most difficult parts of the production. In addition to conventional methods, new instruments and tests have been developed to predict the maturity of crops that help determining the optimum harvest time, although the maturity assessment differs according to crop variety.

At optimum maturity, physiological changes in several vegetables take place very rapidly. Thus, the determination of optimum harvesting time is critical. Some vegetables such as green peas and sweet corn only have a short period during which they are of prime quality. If harvesting is delayed beyond this point, quality deteriorates and the crop may quickly become unacceptable. Most of the vegetables are subjected to bruising during harvesting.

Vegetables at peak flavour and texture are used for freezing. Postharvest delays in handling vegetables are known to produce deterioration in flavour, texture, colour, and nutrients. Therefore, the delays between harvest and processing should be reduced to retain fresh quality prior to freezing. Cooling vegetables by cold water, air blasting, or ice will often reduce the rate of postharvest losses sufficiently, providing extra hours of high quality retention for transporting raw material to considerable distances from the field to the processing plant.

Blanching is the exposure of the vegetables to boiling water or steam for a brief period of time to inactivate enzymes. Practically every vegetable (except herbs and green peppers) needs to be blanched and promptly cooled prior to freezing, since heating slows or stops the enzyme action, which causes vegetables to grow and mature. After maturation, however, enzymes can cause loss in quality, flavour, colour, texture, and nutrients. If vegetables are not heated sufficiently, the enzymes will continue to be active during frozen storage and may cause the vegetables to toughen or develop off-flavours and colours. Blanching also causes wilting or softening of vegetables, making them easier to pack. It destroys some bacteria and helps remove any surface dirt.

Blanching in hot water at 70 to 105 °C has been associated with the destruction of enzyme activity. Blanching is usually carried out between 75 and 95 °C for 1 to 10 minutes, depending on the size of individual vegetable pieces. Blanched vegetables should be promptly cooled down to control and minimize the degradation of soluble and heat-labile nutrients. The enzymes used as indicators of effectiveness of the blanching treatment are peroxidase, catalase, and more recently lipoxygenase. Peroxidase inactivation is commonly used in vegetable processing, since peroxidase is easily detected and is the most heat stable of these enzymes.

Vegetables can be blanched in hot water, steam, and in the microwave. Hot water blanching is the most common way of processing vegetables. For water blanching, vegetables are put in a basket and then placed in a kettle of boiling water covered with a lid. Timing begins immediately. Steam blanching takes longer than the water method, but helps retain water-soluble nutrients such as water-soluble vitamins. For steam blanching, a single layer of vegetables is placed on a rack or in a basket at 3-5 cm above water boiling in a kettle. A tightly fitted lid is placed on the kettle and timing is started. Microwave blanching is usually recommended for small quantities of vegetables prior to freezing. Due to the non-uniform heating disadvantage of microwaves, research is still being conducted to obtain better results with microwave blanching.

There are several factors to consider in packaging frozen vegetables, which include protection from atmospheric oxygen, prevention of moisture loss, retention of flavour, and rate of heat transfer through the package. There are two basic packing methods recommended for frozen vegetables: dry pack and tray pack. In the dry pack method, the blanched and drained vegetables are put into meal-sized freezer bags and packed tightly to cut down on the amount of air in the package. Proper headspace (approximately 2 cm) is left at the top of rigid containers before closing. For freezer bags, the headspace is larger. Provision for headspace is not necessary for foods such as broccoli, asparagus, and brussels sprouts, as they do not pack tightly in containers.

In the tray pack method, chilled, well-drained vegetables are placed in a single layer on shallow trays or pans. Trays are placed in a freezer until the vegetables become firm, then removed. Vegetables are filled into containers. Tray-packed foods do not freeze in a block but remain loosely distributed so that the amount needed can be poured from the container and the package reclosed.

REFERENCES

Archer, G.P., Kennedy, C. J. and Wilson, A. J., "Towards predictive microbiology in frozen food systems-a framework for understanding microbial population dynamics in frozen structures and in freeze-thaw cycles", *International Journal of Food Science and Technology*, 30, pp. 711, 1995.

Blanshard, J. M.V. and Franks, F., *Ice crystallization and its control in frozen-food systems*, Food Structure and behavior, Academic Press, London, UK, pp. 51-65, 1987.

Brady, P.L., *Freezing fruit*, UA Cooperative Extension Service, University of Arkansas, USA, 2002.

Cano, P., Marin, M.A. and Fuster, C., "Freezing of banana slices, Influence of maturity level and thermal treatment prior to freezing", *Journal of Food Science*, 55, 4, pp. 1070, 1990.

Chen, C., "Thermodynamic analysis of freezing and thawing of foods enthalpy and apparent specific heat", *Journal of Food Science*, 50, 1985.

8

Dairy Process Engineering

The dairy industry is divided into two main production areas:

- the primary production of milk on farms-the keeping of cows (and other animals such as goats, sheep etc.) for the production of milk for human consumption;
- the processing of milk-with the objective of extending its saleable life. This objective is typically achieved by
  - heat treatment to ensure that milk is safe for human consumption and has an extended keeping quality, and
  - preparing a variety of dairy products in a semi-dehydrated or dehydrated form (butter, hard cheese and milk powders), which can be stored.

Dairy processing occurs worldwide; however the structure of the industry varies from country to country. In less developed countries, milk is generally sold directly to the public, but in major milk producing countries most milk is sold on a wholesale basis. In Ireland and Australia, for example, many of the large-scale processors are owned by the farmers as co-operatives, while in the United States individual contracts are agreed between farmers and processors. Dairy processing industries in the major dairy producing countries have undergone rationalisation, with a trend towards fewer but larger plants operated by fewer people. As a result, in the United States, Europe, Australia and New Zealand most dairy processing plants are quite large.

Plants producing market milk and products with short shelf life, such as yogurts, creams and soft cheeses, tend to be located on the fringe of urban centres close to consumer markets. Plants manufacturing items with longer shelf life, such as butter, milk powders, cheese and whey powders, tend to be located in rural areas closer to the milk supply. The general tendency worldwide, is towards large processing plants specialising in a limited range of products. There are exceptions, however. In eastern Europe for example, due to the former supply-driven concept of the market, it is still very common for 'city' processing plants to be large multi-product plants producing a wide range of products.

The general trend towards large processing plants has provided companies with the opportunity to acquire bigger, more automated and more efficient equipment. This technological development has, however, tended to increase environmental loadings in some areas due to the requirement for long-distance distribution. Basic dairy processes have changed little in the past decade. Specialised processes such as ultrafiltration (UF), and modern drying processes, have increased the opportunity for the recovery of milk solids that were formerly discharged.

Process

Milk Production

The processes taking place at a typical milk plant include:

- receipt and filtration/clarification of the raw milk;
- separation of all or part of the milk fat;
- pasteurisation;
- homogenisation;
- deodorisation (if required);
- further product-specific processing;
- packaging and storage, including cold storage for perishable products;
- distribution of final products.

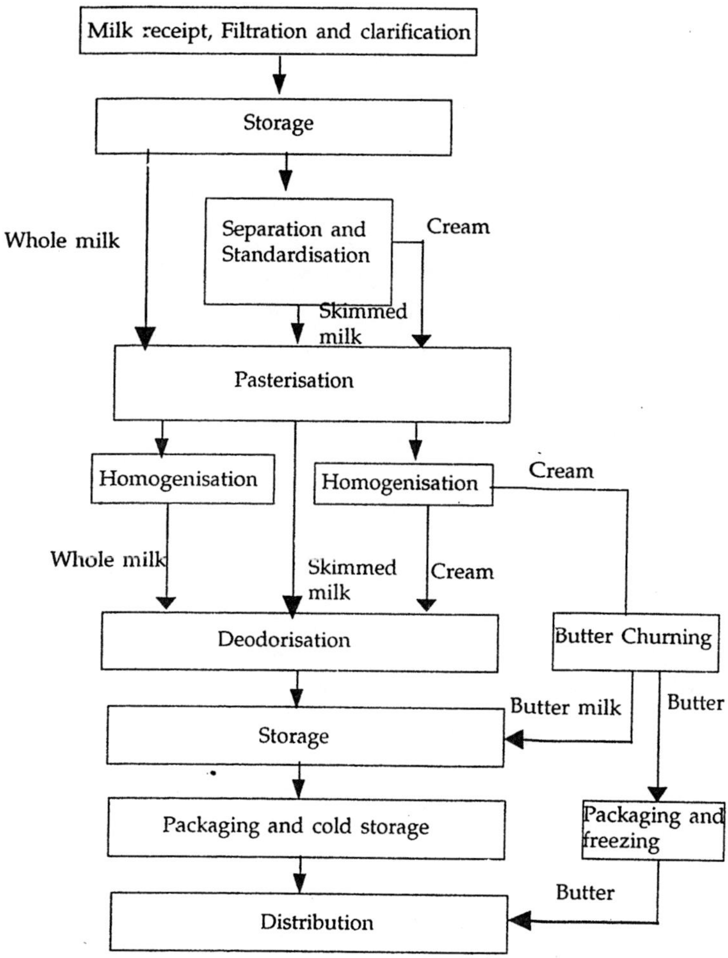

Whole milk Cream Buttermilk Butter Semi-skimmed milk
Skimmed milk

Figure 1. Flow diagram for processes occurring at a typical milk plant

Figure 1 is a flow diagram outlining the basic steps in the production of whole milk, semi-skimmed milk and skimmed milk, cream, butter and buttermilk. In such plants, yogurts and other cultured products may also be produced from whole milk and skimmed milk.

Butter Production

The butter-making process, whether by batch or continuous methods, consists of the following steps:

- preparation of the cream;
- destabilisation and breakdown of the fat and water emulsion;
- aggregation and concentration of the fat particles;
- formation of a stable emulsion;
- packaging and storage;
- distribution.

Figure 2 is a flow diagram outlining the basic processing system for a butter-making plant. Milk destined for butter making must not be homogenised, because the cream must remain in a separate phase. After separation, cream to be used for butter making is heat treated and cooled under conditions that facilitate good whipping and churning.

It may then be ripened with a culture that increases the content of diacetyl, the compound responsible for the flavour of butter. Alternatively, culture inoculation may take place during churning. Butter which is flavour enhanced using this process is termed lactic, ripened or cultured butter. This process is very common in continental European countries. Although the product is claimed to have a superior flavour, the storage life is limited. Butter made without the addition of a culture is called sweet cream butter. Most butter made in the English-speaking world is of this nature.

Both cultured and sweet cream butter can be produced with or without the addition of salt. The presence of salt affects both the flavour and the keeping quality. Butter is usually packaged in bulk quantities (25 kg) for long-term storage and then re-packed into marketable portions (usually 250 g or 500 g, and single-serve packs of 10-15 g).

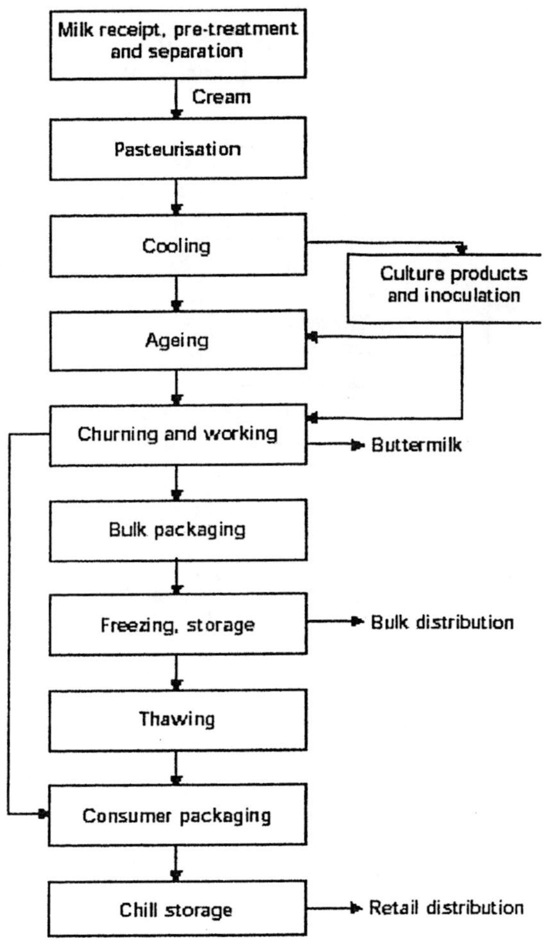

Figure 2. Flow diagram for a typical butter-making plant

Butter may also be packed in internally lacquered cans, for special markets such as the tropics and the Middle East.

Cheese Production

Virtually all cheese is made by coagulating milk protein (casein) in a manner that traps milk solids and milk fat into a curd matrix. This curd matrix is then consolidated to express the liquid fraction, cheese whey. Cheese whey contains those milk solids which are not held in the curd mass, in particular most of the milk sugar (lactose) and a number of soluble proteins.

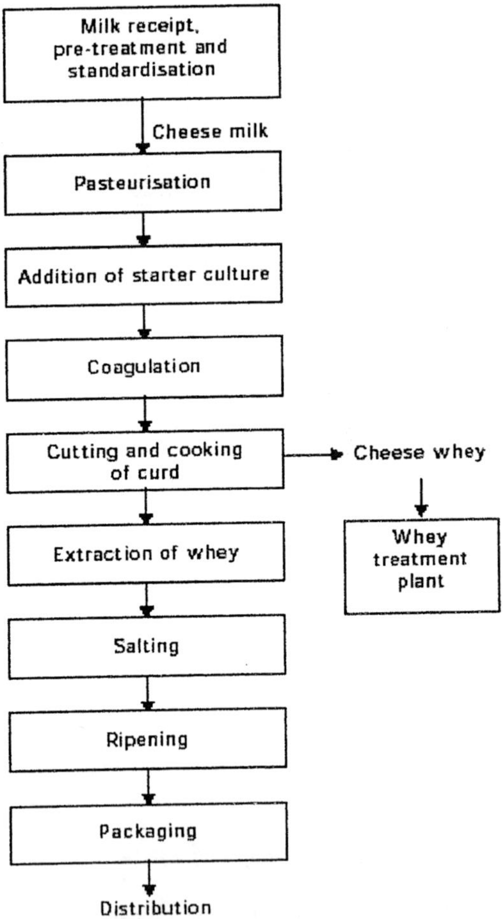

Figure 3. Flow diagram for a typical cheese plant

Figure 3 outlines the basic processes in a cheese-making plant. All cheese-making processes involve some or all of these steps.

Production of Milk Powder

Milk used for making milk powder, whether it be whole or skim milk, is not pasteurised before use. The milk is preheated in tubular heat exchangers before being dried. The preheating temperature depends on the season (which affects the stability of the protein in the milk) and on the characteristics desired for the final powder product.

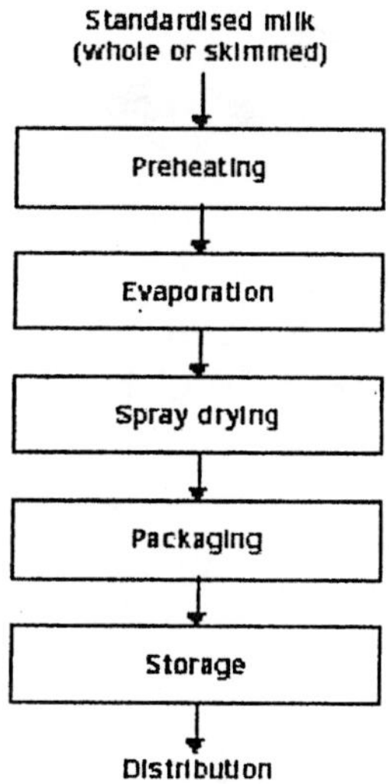

Figure 4 Flow diagram for a typical milk drying planat

The preheated milk is fed to an evaporator to increase the concentration of total solids. The solids concentration that can be reached depends on the efficiency of the equipment and the amount of heat that can be applied without unduly degrading the milk protein.

The milk concentrate is then pumped to the atomiser of a drying chamber. In the drying chamber the milk is dispersed as a fine fog-like mist into a rapidly moving hot air stream, which causes the individual mist droplets to instantly evaporate. Milk powder falls to the bottom of the chamber, from where it is removed. Finer milk powder particles are carried out of the chamber along with the hot air stream and collected in cyclone separators. Milk powders are normally packed and distributed in bulk containers or in 25 kg paper packaging systems. Products sold to the consumer market are normally packaged in cans under nitrogen. This packaging system improves the keeping quality, especially for products with high fat content.

ENVIRONMENTAL IMPACT OF MILK PRODUCTION

Primary Production Impacts

The main environmental issues associated with dairy farming are:

- the generation of solid manure and manure slurries, which may pollute surface water and groundwater;
- the use of chemical fertilisers and pesticides in the production of pastures and fodder crops, which may pollute surface water and groundwater;
- the contamination of milk with pesticides, antibiotics and other chemical residues.

In most cases, solid manure is applied to pastures and cultivated land.

The extent of application, however, may be restricted in some regions. Dairy effluent and slurries are generally held in some form of lagoon to allow sedimentation and biological degradation before they are irrigated onto land. Sludge generated from biological treatment of the dairy effluent can also be applied to pastures, as long as it is within the allowable concentrations for specified pollutants, as prescribed by regulations. Sludge can also be used in the production of methane-rich biogas, which can then be used to supplement energy supplies.

Manure waste represents a valuable source of nutrients. However improper storage and land application of manure and slurries can result in serious pollution of surface waters and groundwater, potentially contaminating drinking water supplies. The extensive use of chemical fertilisers containing high levels of nitrogen has resulted in pollution of the groundwater and surface waters in many countries.

Nitrite in drinking water is known to be carcinogenic, and nitrite levels in drinking water that exceed 25-50 mg/L have been linked to cyanosis in newborn infants ('blue babies'). Compounds containing nitrogen and phosphorus, if discharged to surface water, can lead to excessive algal growth (eutrophication). This results in depleted dissolved oxygen levels in the water, thereby causing the death of fish and other aquatic species. In sensitive areas, therefore, the rate and manner of application of chemical fertilisers are critical.

The use of pesticides has been recognised as an environmental concern for many agricultural activities. Toxic pesticides, some of which biodegrade very slowly, can accumulate in body tissues and are harmful to ecosystems and to human health. Pesticides can end up in agricultural products, groundwater and surface waters, and in extreme cases can enter the human food chain through milk.

For the past few decades, the contamination of milk with antibiotics has been an issue of concern. This is due to the overuse of antibiotics for treatment of cattle diseases, particularly mastitis. It has been brought under control in most countries with developed dairy industries, through strict limitations on the use of antibiotics, regular testing of milk for antibiotic residues, rigorous enforcement of regulations, and education. In some countries, considerable attention has also been paid to the screening of milk supplies for traces of radioactivity, and most countries now apply acceptance limits for raw and imported milk products. Even the slightest levels of contamination in milk can be serious, because pollutants are concentrated in the processing process.

Dairy Processing Impacts

As for many other food processing operations, the main environmental impacts associated with all dairy processing activities are the high consumption of water, the discharge of effluent with high organic loads and the consumption of energy. Noise, odour and solid wastes may also be concerns for some plants. Dairy processing characteristically requires very large quantities of fresh water. Water is used primarily for cleaning process equipment and work areas to maintain hygiene standards. The dominant environmental problem caused by dairy processing is the discharge of large quantities of liquid effluent. Dairy processing effluents generally exhibit the following properties:

— high organic load due to the presence of milk components;
— fluctuations in pH due to the presence of caustic and acidic cleaning agents and other chemicals;
— high levels of nitrogen and phosphorus;
— fluctuations in temperature.

If whey from the cheese-making process is not used as a by-product and discharged along with other wastewaters, the organic load of the resulting effluent is further increased, exacerbating the environmental problems that can result.

In order to understand the environmental impact of dairy processing effluent, it is useful to briefly consider the nature of milk. Milk is a complex biological fluid that consists of water, milk fat, a number of proteins (both in suspension and in solution), milk sugar (lactose) and mineral salts. Dairy products contain all or some of the milk constituents and, depending on the nature and type of product and the method of manufacturing, may also contain sugar, salts (e.g. sodium chloride), flavours, emulsifiers and stabilisers.

For plants located near urban areas, effluent is often discharged to municipal sewage treatment systems. For some municipalities, the effluent from local dairy processing plants can represent a significant load on sewage treatment plants. In extreme cases, the organic load of waste milk solids entering a sewage system may well exceed that of the township's domestic waste, overloading the system. In rural areas, dairy processing effluent may also be irrigated to land. If not managed correctly, dissolved salts contained in the effluent can adversely affect soil structure and cause salinity. Contaminants in the effluent can also leach into underlying groundwater and affect its quality.

In some locations, effluent may be discharged directly into water bodies. However this is generally discouraged as it can have a very negative impact on water quality due to the high levels of organic matter and resultant depletion of oxygen levels. Electricity is used for the operation of machinery, refrigeration, ventilation, lighting and the production of compressed air. Like water consumption, the use of energy for cooling and refrigeration is important for ensuring good keeping quality of dairy products and storage temperatures are often specified by regulation. Thermal energy, in the form of steam, is used for heating and cleaning. As well as depleting fossil fuel resources, the consumption of energy causes air pollution and greenhouse gas emissions, which have been linked to global warming.

Dairy products such as milk, cream and yogurt are typically packed in plastic-lined paperboard cartons, plastic bottles and cups, plastic bags or reusable glass bottles. Other products, such as butter and cheese, are wrapped in foil, plastic film or small plastic containers. Milk powders are commonly packaged in multi-layer kraft paper sacs or tinned steel cans, and some other products, such as condensed milks, are commonly packed in cans.

Breakages and packaging mistakes cannot be totally avoided. Improperly packaged dairy product can often be returned for reprocessing; however the packaging material is generally discarded. Emissions to air from dairy processing plants are caused by the high levels of energy consumption necessary for production. Steam, which is used for heat treatment processes (pasteurisation, sterilisation, drying etc.) is generally produced in on-site boilers, and electricity used for cooling and equipment operation is purchased from the grid. Air pollutants, including oxides of nitrogen and sulphur and suspended particulate matter, are formed from the combustion of fossil fuels, which are used to produce both these energy sources.

In addition, discharges of milk powder from the exhausts of spray drying equipment can be deposited on surrounding surfaces. When wet these deposits become acidic and can, in extreme cases, cause corrosion. For operations that use refrigeration systems based on chlorofluorocarbons (CFCs), the fugitive loss of these gases to the atmosphere is an important environmental consideration, since CFCs are recognised to be a cause of ozone depletion in the atmosphere. For such operations, the replacement of CFC-based systems with non- or reduced-CFC systems is thus an important issue.

Some processes, such as the production of dried casein, require the use of hammer mills to grind the product. The constant noise generated by this equipment has been known to be a nuisance in surrounding residential areas. The use of steam injection for heat treatment of milk and for the creation of reduced pressure in evaporation processes also causes high noise levels. A substantial traffic load in the immediate vicinity of a dairy plant is generally unavoidable due to the regular delivery of milk (which may be on a 24-hour basis), deliveries of packaging and the regular shipment of products. Noise problems should be taken into consideration when determining plant location.

Hazardous wastes consist of oily sludge from gearboxes of moving machines, laboratory waste, cooling agents, oily paper filters, batteries, paint cans etc. At present, in western Europe some of these materials are collected by waste companies. While some waste is incinerated, much is simply dumped.

Environmental Indicators for Production

Environmental indicators are important for assessing Cleaner Production opportunities and for assessing the environmental performance of one dairy processing operation relative to another. They provide an indication of resource consumption and waste generation per unit of production. Figure 5 is a generic flowchart of the overall process including resource inputs and waste outputs.

As with most food processing operations, water is used extensively for cleaning and sanitising plant and equipment to maintain food hygiene standards. The areas of water consumption within a dairy processing plant, and gives an indication of the extent to which each area contributes to overall water use. Due to the higher costs of water and effluent disposal that have now been imposed in some countries to reflect environmental costs, considerable reduction in water consumption has been achieved over the past few decades in the dairy processing industry. These improvements are attributed to developments in process control and cleaning practices.

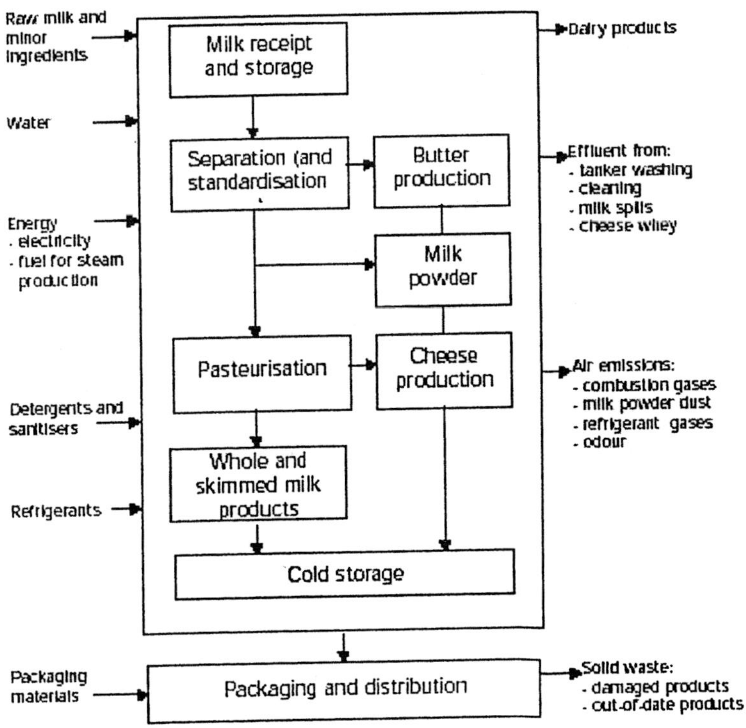

Figure 5. Inputs and outputs of a dairy

At modern dairy processing plants, a water consumption rate of 1.3-2.5 litres water/kg of milk intake is typical, however 0.8-

1.0 litres water/kg of milk intake is possible. To achieve such low consumption requires not only advanced equipment, but also very good housekeeping and awareness among both employees and management.

Dairy processing effluent contains predominantly milk and milk products which have been lost from the process, as well as detergents and acidic and caustic cleaning agents. The constituents present in dairy effluent are milk fat, protein, lactose and lactic acid, as well as sodium, potassium, calcium and chloride. Milk loss to the effluent stream can amount to 0.5-2.5% of the incoming milk, but can be as high as 3-4%. A major contributing factor to a dairy plant's effluent load is the cumulative effect of minor and, on occasions, major losses of milk. These losses can occur, for example, when pipework is uncoupled during tank transfers or equipment is being rinsed.

The organic pollutant content of dairy effluent is commonly expressed as the 5-day biochemical oxygen demand (BOD5) or as chemical oxygen demand (COD). One litre of whole milk is equivalent to approximately 110,000 mg BOD5 or 210,000 mg COD. Concentrations of COD in dairy processing effluents vary widely, from 180 to 23,000 mg/L. Low values are associated with milk receipt operations and high values reflect the presence of whey from the production of cheese. A typical COD concentration for effluent from a dairy plant is about 4000 mg/L. This implies that 4% of the milk solids received into the plant is lost to the effluent stream, given that the COD of whole milk is 210,000 mg/L and that effluent COD loads have been estimated to be approximately 8.4 kg/m3 milk intake.

A Danish survey found that effluent loads from dairy processing plants depend, to some extent, on the type of product being produced. The scale of the operation and whether a plant uses batch or continuous processes also have a major influence, particularly for cleaning. This is because small batch processes requires more frequent cleaning. The tendency within the industry towards larger plants is thus favourable in terms of pollutant loading per unit of production.

Due to the traditional payment system for raw milk (which is based on the mass or volume delivered plus a separate price or premium for the weight of milk fat), the dairy processing industry has always tried to minimise loss of milk fat. In many countries the payment system now recognises the value of the non-fat milk components. Systems that control the loss of both fat and protein are now common in the industrialised world, but less so in the developing world.

The disposal of whey produced during cheese production has always been a major problem in the dairy industry. Whey is the liquid remaining after the recovery of the curds formed by the action of enzymes on milk. It comprises 80-90% of the total volume of milk used in the cheese making process and contains more than half the solids from the original whole milk, including 20% of the protein and most of the lactose. It has a very high organic content, with a COD of approximately 60,000 mg/L. Only in the past two decades have technological advances made it economically possible to recover soluble proteins from cheese whey and, to some extent, to recover value from the lactose.

Most dairies are aware that fat and protein losses increase the organic load of the effluent stream and, even in the developing world, the use of grease traps has been common for some decades. Many companies, however, do not take any action to reduce the organic pollution from other milk components. It is becoming more common for dairy companies to be forced by legal or economic pressures to reduce the amount and concentration of pollutants in their effluent streams. Therefore, at most sites, wastewater treatment or at least pretreatment is necessary to reduce the organic loading to a level that causes minimal environmental damage and does not constitute a health risk. The minimum pretreatment is usually neutralisation of pH, solids sedimentation and fat removal.

Energy Consumption Process

Energy is used at dairy processing plants for running electric motors on process equipment, for heating, evaporating and drying, for cooling and refrigeration, and for the generation of compressed air. Approximately 80% of a plant's energy needs is met by the combustion of fossil fuel (gas, oil etc.) to generate steam and hot water for evaporative and heating processes. The remaining 20% or so is met by electricity for running electric motors, refrigeration and lighting.

The energy consumed depends on the range of products being produced. Processes which involve the concentration and drying of milk, whey or buttermilk for example, are very energy intensive. The production of market milk at the other extreme involves only some heat treatment and packaging, and therefore requires considerably less energy.

Energy consumption will also depend on the age and scale of a plant as well as the level of automation. Plants producing powdered milk exhibit a wide range of energy efficiencies, depending on the type of evaporation and drying processes that are used. Energy consumption depends on the number of evaporation effects (the number of evaporation units that are used in series) and the efficiency of the powder dryer. Substantial increases in electricity use have resulted from the trend towards automated plant with associated pumping costs and larger evaporators as well as an increase in refrigeration requirements. High consumption of electricity can also be due to the use of old motors, excessive lighting or possibly a lack of power factor correction.

Cleaner Production Opportunities

Dairy processing typically consumes large quantities of water and energy and discharges significant loads of organic matter in the effluent stream. For this reason, Cleaner Production opportunities described in this guide focus on reducing the consumption of resources (water and energy), increasing production yields and reducing the volume and organic load of effluent discharges.

At the larger production scales, dairy processing has become an extremely automated process and resource efficiency relies, to a large extent, on the efficiency of plant and equipment, the control systems that are used to operate them and the technologies used to recover resources. As a result many Cleaner Production opportunities lie in the selection, design and efficient operation of process equipment. Operator practices also have an impact on plant performance, for example in the areas of milk delivery, plant maintenance and cleaning operations.

Many food processors that undertake Cleaner Production projects find that significant environmental improvement and cost savings can be derived from simple modification to housekeeping procedures and maintenance programmes.

Water is used extensively in dairy processing, so water saving measures are very common Cleaner Production opportunities in this industry. The first step is to analyse water use patterns carefully, by installing water meters and regularly recording water consumption. Water consumption data should be collected during production hours, especially during periods of cleaning. Some data should also be collected outside normal working hours to identify leaks and other areas of unnecessary wastage.

The next step is to undertake a survey of all process area and ancillary operations to identify wasteful practices. Examples might be hoses left running when not in use, CIP cleaning processes using more water than necessary, etc. Installing automatic shut-off equipment and restrictors could prevent such wasteful practices. Automatic control of water use is preferable to relying on operators to manually turn water off. Once wasteful practices have been addressed, water use for essential process functions can be investigated. It can be difficult to establish the minimum consumption rate necessary to maintain process operations and food hygiene standards.

The optimum rate can be determined only by investigating each process in detail and undertaking trials. Such investigations should be carried out collaboratively by production managers, food quality and safety representatives and operations staff. When an optimum usage rate been agreed upon, measures should be taken to set the supply at the specified rate and remove manual control. Once water use for essential operations has been optimised, water reuse can be considered.

Table 1. Checklist of ideas for reducing pollutant loads in effluent

- — Ensure that vessels and pipes are drained completely and using pigs and plugs to remove product residues before cleaning;
- — Use level controls and automatic shut-off systems to avoid spills from vessels and tanker emptying;
- — Collect spills of solid materials (cheese curd and powders) for reprocessing or use as stock feed;
- — Fit drains with screens and/or traps to prevent solid materials entering the effluent system;
- — Install in-line optical sensors and diverters to distinguish between product and water and minimise losses of both;
- — Install and maintain level controls and automatic shut-off systems on tanks to avoid overfilling;
- — Use dry cleaning techniques where possible, by scraping vessels before cleaning or pre-cleaning with air guns;
- — Use starch plugs or pigs to recover product from pipes before internally cleaning tanks.

Wastewaters that are only slightly contaminated could be used in other areas. For example, final rinse waters could be used as the initial rinses for subsequent cleaning activities, or evaporator condensate could be reused as cooling water or as boiler feed water. Wastewater reuse should not compromise product quality and hygiene, and reuse systems should be carefully installed so that reused wastewater lines cannot be mistaken for fresh water lines, and each case should be approved by the food safety officer.

Cleaner Production efforts in relation to effluent generation should focus on reducing the pollutant load of the effluent. The volume of effluent generated is also an important issue. However this aspect is linked closely to water consumption, therefore efforts to reduce water consumption will also result in reduced effluent generation. Opportunities for reducing the pollutant load of dairy plant effluent focus on avoiding the loss of raw materials and products to the effluent stream. This means avoiding spills, capturing materials before they enter drains and limiting the extent to which water comes into contact with product residues. Improvements to cleaning practices are therefore an area where the most gains can be made. Table 1 contains a checklist of common ideas for reducing effluent loads.

Energy is an area where substantial savings can be made almost immediately with no capital investment, through simple housekeeping and plant optimization efforts. Substantial saving are possible through improved housekeeping and the fine tuning of existing processes and additional savings are possible through the use of more energy-efficient equipment and heat recovery systems.

In addition to reducing a plant's demand for energy, there are opportunities for using more environmentally benign sources of energy. Opportunities include replacing fuel oil or coal with cleaner fuels, such as natural gas, purchasing electricity produced from renewable sources, or co-generation of electricity and heat on site. For some plants it may also be feasible to recover methane from the anaerobic digestion of highstrength effluent streams to supplement fuel supplies.

Table 2. Checklist of energy saving ideas

— Implement switch-off programmes and installing sensors to turn off or power down lights and equipment when not in use;
— Improve insulation on heating or cooling systems and pipework;
— Favour more energy-efficient equipment;
— Improve maintenance to optimise energy efficiency of equipment;
— Maintain optimal combustion efficiencies on steam and hot water boilers;
— Eliminate steam leaks;
— Capture low-grade energy for use elsewhere in the operation.

Milk Production

Storage of Milk

Raw milk is generally received at processing plants in milk tankers. Some smaller plants may also receive milk in 25-50 Laluminium or steel cans or, in some less developed countries, in plastic barrels. Depending on the structure and traditions of the primary production sector, milk may be collected directly from the farms or from central collection facilities. Farmers producing only small amounts of milk normally deliver their milk to central collection facilities.

At the central collection facilities, operators measure the quantity of milk and the fat content. The milk is then filtered and/or clarified using centrifuges to remove dirt particles as well as udder and blood cells. The milk is then cooled using a plate cooler and pumped to insulated or chilled storage vessels, where it is stored until required for production. Empty tankers are cleaned in a wash bay ready for the next trip. They are first rinsed internally with cold water and then cleaned with the aid of detergents or a caustic solution. To avoid build-up of milk scale, it is then necessary to rinse the inside of the tank with a nitric acid wash. Tankers may also be washed on the outside with a cold water spray. Until required for processing, milk is stored in bulk milk vats or in insulated vessels or vessels fitted with water jackets.

Water is consumed for rinsing the tanker and cleaning and sanitising the transfer lines and storage vessels. The resulting effluent from rinsing and cleaning can contain milk spilt when tanker hoses are disconnected. This would contribute to the organic load of the effluent stream. Solid waste is generated from milk clarification and consists mostly of dirt, cells from the cows' udders, blood corpuscles and bacteria. If this is discharged into the effluent stream, high organic loads and associated downstream problems can result. Cleaner Production opportunities in this area focus on reducing the amount of milk that is lost to the effluent stream and reducing the amount of water used for cleaning. Ways of achieving this include:

- — avoiding milk spillage when disconnecting pipes and hoses;
- — ensuring that vessels and hoses are drained before disconnection;
- — providing appropriate facilities to collect spills;
- — identifying and marking all pipeline to avoid wrong connections that would result in unwanted mixing of products;
- — installing pipes with a slight gradient to make them self-draining;
- — equipping tanks with level controls to prevent overflow;
- — making certain that solid discharges from the centrifugal separator are collected for proper disposal and not discharged to the sewer;
- — using 'clean-in-place' (CIP) systems for internal cleaning of tankers and milk storage vessels, thus improving the effectiveness of cleaning and sterilisation and reducing detergent consumption;
- — improving cleaning regimes and training staff;
- — installing trigger nozzles on hoses for cleaning;
- — reusing final rinse waters for the initial rinses in CIP operations;
- — collecting wastewaters from initial rinses and returning them to the dairy farm for watering cattle.

Separation and Standardisation

Dairies that produce cream and/or butter separate fat from the raw milk. Separation takes place in a centrifuge which divides the milk into cream with about 40% fat and skimmed milk with only about 0.5% fat. The skimmed milk and cream are stored and pasteurised separately. Most dairies standardise all milk, to ensure that their products have a consistent composition. In some cases, products may need to meet certain product specifications in relation to fat content. These specifications vary from one country to the next.

However in general, whole milk must contain around 3.5-4.2% fat, semi-skimmed milk around 1.3-1.5% and skimmed milk around 0.5%. Standardisation is achieved by the controlled remixing of cream with skimmed milk, and is common both in cheese plants and in the production of milk powders.

As in other aspects of dairy processing, water is consumed for rinsing and cleaning of process equipment, resulting in the generation of wastewaters containing milk solids and cleaning agents. The centrifugal separators generate a sludge material, which consists of udder and blood cells and bacteria contained in the raw milk. For standard separators the sludge is removed manually during the cleaning phase, while in the case of self-cleaning centrifuges it is discharged automatically. If the sludge is discharged to the sewer along with the effluent stream, it greatly increases the organic load of the effluent. Cleaner Production opportunities specific to this area are related to reducing the generation of separator sludge and optimising its collection and disposal. Ways of achieving this include:

- reducing the frequency with which centrifugal separators are cleaned, by improving milk filtration at the receiving stage or by clarification of the raw milk;
- collecting the sludge and disposing of it along with other waste solids.

Also of importance is the optimization of cleaning processes, to make them water and energy efficient.

Pasteurisation and Homogenisation

In large plants, milk is pasteurised in continuous flow pasteurisers, whereas some smaller dairies may use batch-type pasteurisers. In batch pasteurisation processes, milk is typically heated to 62.8-65.6°C for 30 minutes, whereas in continuous pasteurisation processes it is heated to 71.7-78.1°C for at least 15 seconds. The time-temperature relationship is usually prescribed by law, as are certain safeguards to ensure that all milk attains the minimum treatment. For both batch and continuous processes, the milk is cooled to below 10°C immediately after heating.

For some products milk is homogenised using a pressure pump, which breaks up the butterfat globules to a size that keeps them in suspension. In continuous pasteurisation processes homogenisation is usually undertaken in conjunction with pasteurisation, since its efficiency is improved if the milk is warm. The main environmental issue associated with pasteurisation and homogenisation is the high levels of energy consumed for the heating and cooling of milk. In addition, water is consumed for rinsing and cleaning of process equipment, resulting in the generation of wastewaters containing milk solids and cleaning agents. In batch pasteurisation, small batches necessitate frequent cleaning, therefore losses of milk and the organic loads in wastewater streams are increased.

Cleaner Production opportunities in this area focus on improving energy efficiency. Ways of achieving this include:

- — replacing batch pasteurisers with a continuous process incorporating plate heat exchanger (PHE) pasteurisers, where feasible. PHE pasteurisers are more energy efficient than batch pasteurisers because the heat from the pasteurised milk can be used to preheat the incoming cold milk (regenerative countercurrent flow);
- — installing new manufacturing equipment, which will result in less waste of milk products than the equipment currently used in many dairies;
- — avoiding stops in continuous processes. The more constant the production, the less milk will be lost, since most waste comes from cleaning of batch process equipment. In the event of upgrades to process equipment, high-volume pasteurising units should be considered;
- — reducing the frequency of cleaning of the pasteuriser. Particularly for small dairies, optimising the size of balance tanks before and after the pasteuriser will allow continuous operation of the pasteuriser and reduce cleaning frequency;

— planning production schedules so that product change-overs coincide with cleaning regimes;
— collecting and recovering the milky wastewater generated at startup of pasteurisation and supplying it to farmers as animal feed.

Also of importance is optimization of cleaning processes, to make them water and energy efficient. To make possible the reprocessing of excess milk returned from the market, dairy plants may wish to consider developing policies which allow for the reprocessing of milk without affecting the quality of the freshly pasteurised product.

The introduction of poorer quality milk into the pasteurisation process can result in milk scale and coagulation problems due to higher acidity. This may cause higher milk losses in the pasteuriser due to the need for more frequent cleaning in order to remove milk scale. These issues should be weighed against the benefits of reprocessing returned milk. The controlled return and reprocessing of milk from the market may require training of sales representatives. Alternatively, penalties could be applied for inappropriate ordering, or bonuses paid for extended periods of no market returns.

Deodorisation Units

Many dairies remove unwanted taints and odours from milk in deodorisation units. In these systems, the odorous substances are drawn-off by injecting steam into the system under vacuum. In situations where the taints and odours are only mild, a vacuum alone may be used. Figure 6 is a flow diagram showing the inputs and outputs for this process.

An environmental issue specific to the deodorisation process is the large volume of water used to operate water seals on the vacuum pump. Water used for the vacuum pump can be recirculated to reduce or eliminate the necessity to discharge it.

Storage and Packaging

Due to the large range of products produced at many dairies, the bulk storage of these products can involve very extensive storage systems, with associated vessels, piping and valves. Milk is packaged or bottled in a number of types of containers, including glass bottles, paper cartons, plastic bottles and plastic pouches.

In most cases, filling of containers is highly automated. After filling, the packaged milk products are usually stored and transported in wire or plastic crates.

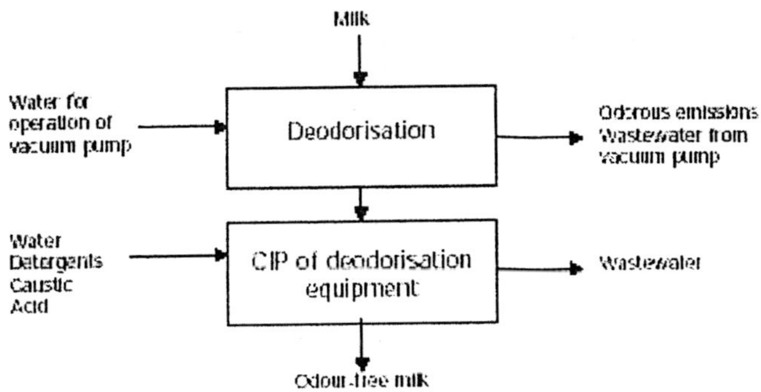

Figure 6. Inputs and outputs for the deodorisation of milk

Finished products are held in refrigerated storage until dispatched to retail outlets. The storage temperature depends on the product, but for milk and fresh dairy products, the optimum temperature is usually <4°C. Refrigerated storage chambers are usually cooled using forced draft evaporators chilled by a primary refrigerant.

A secondary refrigerant such as ice water, brine or glycol recirculated in a closed circuit cooling system is also sometimes used. Door openings are usually sealed with rubber swing doors and/or air curtains when open. Figure 7. is a flow diagram showing the inputs and outputs for this process.

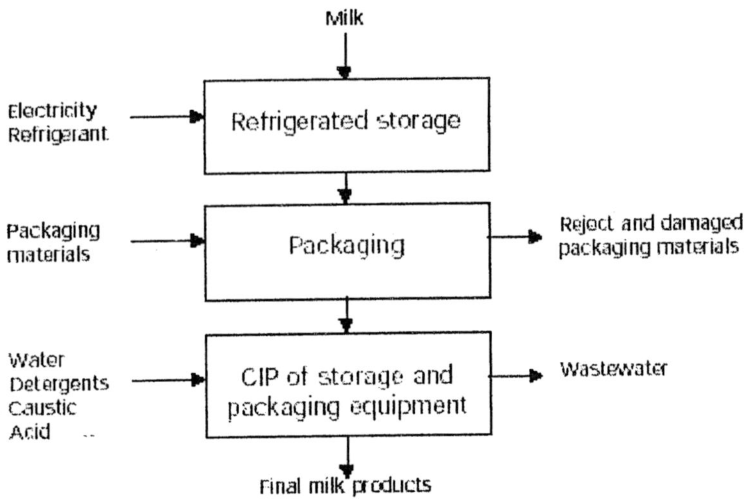

Figure 7. Inputs and outputs for storage and packaging of milk products

The main environmental issues associated with the storage and packaging operations are the loss of milk products from spills and packaging mistakes, generation of wastewater from cleaning processes and energy consumed for refrigerated storage. However the choice of packaging materials is becoming an increasingly important issue.

Milk products can be lost to the wastewater stream during start-up and shut-down, from residues remaining in storage vessels and from the initial cold water rinses of packaging and storage equipment. Milk products may also be lost due to breakage of packaging material.

Considerable work has been undertaken to determine the most suitable form of packaging in terms of overall environmental impacts. Although glass bottles can be cleaned and recycled (thereby creating minimal solid waste), cleaning them consumes water and energy. Glass recycling systems require large capital investments and involve high running costs since the bottles must be collected, then transported and cleaned. Glass bottles can also be inconvenient for consumers because they are heavier and more fragile than cartons.

Cartons, on the other hand, create solid waste that must be transported and disposed of. Solid waste can be disposed of in a landfill, incinerated, or composted. All of these disposal alternatives have environmental impacts, including the generation of leachate from landfills and air pollution from incineration.

Cleaner Production opportunities in this area focus on improving the energy efficiency of refrigeration systems and optimising CIP processes to reduce both water use and the organic load discharged into the effluent stream. Ways of achieving this include:

— clearing milk residues from the pipes using compressed air before the first rinse;

— collecting the more highly concentrated milk wastewater at startup and shutdown for use as animal feed;

— optimising the accuracy of filling operations. This will not only result in improved efficiency, but will also reduced potential for waste and spillage. Minor variations in filling performance can have significant cumulative effects particularly for small unit fill quantities;

— improving procedures for recovering milk from wrongly filled containers;

— emptying and collecting product from wrongly filled containers for use as animal feed;

— reducing energy consumption through improved insulation, closing of doors to cold areas, good maintenance of room coolers and regular defrosting;

— using direct ammonia-based cooling systems instead of CFC-based systems.

The primary objective of butter making is to conserve the fatty portion of the milk in a form that can be used at a later date. It is essentially a dehydration process, in which the majority of the aqueous phase is removed and the remainder is emulsified into the fat. Milk is an emulsion of milk fat in water, whereas butter is an emulsion of water in milk fat. Butter production involves the conversion from one state to the other.

The evolution of the butter-making process has progressed from the use of skins and gourds for churning, through to the use of wooden-barrelled butter churns, which have since been exchanged for stainless steel churns. Although the development of the continuous process in the 1950s led to the replacement of the batch process in most industrial plants, the batch or churn process may still be used in smaller dairies. In batch processes, prepared cream is agitated in a specially designed vessel (butter churn) until phase inversion occurs and the fat 'breaks' from the cream in the form of butter grains.

The surrounding liquid-the buttermilk-is then decanted off. The butter grains are washed in fresh chilled water, salted (if required) and worked by a shearing process to produce a homogeneous mass with a controlled moisture content. In the more common continuous process, phase inversion of the cream, working of the butter, the addition of salt and moisture control take place in cylindrical, rotating chambers which progressively lead the butter mass to blending augers and final extrusion. The continuous process reduces the amount of waste generated by the process by eliminating the butter grain washing step and also by making use of an internal mechanical system for continuous recovery of butter 'fines'.

Cream Treatment

Pasteurisation of the cream for making cultured butter is normally carried out at temperatures of up to 110°C. The cream may be subjected to vacuum treatment during cooling in order to improve its spreadability. In the production of ripened butter, the cream is cooled, inoculated with a culture and ripened. After a ripening period of 12-18 hours at 20°C, the cream is cooled to below 10°C.

The cream treatment process has received considerable attention over many years because it affects the quality of the final product. The quality of the fat before it is churned affects product losses from the process. The optimum temperature for ageing the cream (allowing all fat to become solid) is generally lower than the temperature required for efficient churning. Cream that is too cold is therefore susceptible to damage, and may result in blocked pipeline and excessive loss. The most effective churning temperature for cream can be achieved by using heat exchangers with a low pressure drop and a minimum temperature differential between the cream and the water. This avoids localised overheating.

The main environmental issue associated with this process is the high organic load in wastewaters generated from rinsing and cleaning the pasteuriser. This can be further exacerbated by the requirement for frequent cleaning, which results in a greater loss of milk solids. Cleaner Production opportunities in this area focus on reducing water use and loss of product. Ways of achieving this include:

- minimising the number of times the pasteuriser is cleaned. Particularly in small butter dairies, optimising the size of balance tanks before and after the pasteuriser will allow it to operate continuously, resulting in less need for cleaning;
- installing modern pasteurising equipment. This will reduce waste of cream in many dairies, because improvements in plate design now give a more gentle and constant heat treatment. This decreases the build-up of overheated solids on heating surfaces. In the event of upgrades to process equipment, high-volume pasteurising units should be considered;
- collecting the more highly concentrated milk wastewater generated when starting up the pasteuriser, for use as animal feed.

The cream enters the butter maker and the fat globules are disrupted under controlled conditions to destabilise the emulsion and agglomerate the milk fat. This is achieved in the first churning cylinder, which is fitted with a beater driven by a variable-speed motor. The beater speed is adjusted to give the desired butter grain size with minimum fat loss in the buttermilk.

To maintain steady butter-making conditions, it is essential that the cream feed rate be constant. This can be achieved by using a balance tank between the ageing silo and the pump. The mixture of butter grains and buttermilk falls from the first cylinder into the back section of a second cylinder, where the grains are consolidated. This second cylinder is a larger, perforated, slowly rotating drum which causes the grains to travel along an inclined rotating screen with a tumbling action, thus assisting their aggregation at the same time as they are drained of buttermilk. The buttermilk is pumped away from below the cylinder.

From the second cylinder, the moist grains of butter fall into the worker compartment which uses contra-rotating augers to compact the grains into a heterogeneous mass, expelling more buttermilk from the grains as they are squeezed together. Compacted butter grains are fed from the auger through a series of alternating perforated plates and impeller blades. These apply shear forces that further consolidate the butter grains and break up the droplets of buttermilk now remaining in the fat matrix.

This forms a dispersed aqueous phase of what is now a waterin-oil emulsion. A second worker compartment, operating under vacuum, may be incorporated to obtain a denser, finer-textured product. A second set of augers removes the butter and forces it through a final set of orifice plates and blades which complete the emulsification before the product is discharged from the butter maker. Figure 8 is a flow diagram showing the inputs and outputs for this process.

Unless the buttermilk is used as a product or as an ingredient in other products, the quantities of buttermilk produced (about 50% of the original cream volume) represent a potential environmental loading. Pollutant discharge is greatest when a continuous butter maker is closed down, due to the loss of the fat remaining in the machine. Cleaner

Production opportunities in this area focus on reducing loss of product. Ways of achieving this include:

- ensuring that the buttermilk is collected separately and hygienically so that it can be used in other processes, such as a base for lowfat spreads;
- collecting all first rinses, and separating the residual fat for use in other processes;
- preventing the build-up of milk scale deposits;
- maintaining butter makers on a regular basis;
- avoiding spills by ensuring that the buttermilk collection facilities are large enough to hold all the liquid.

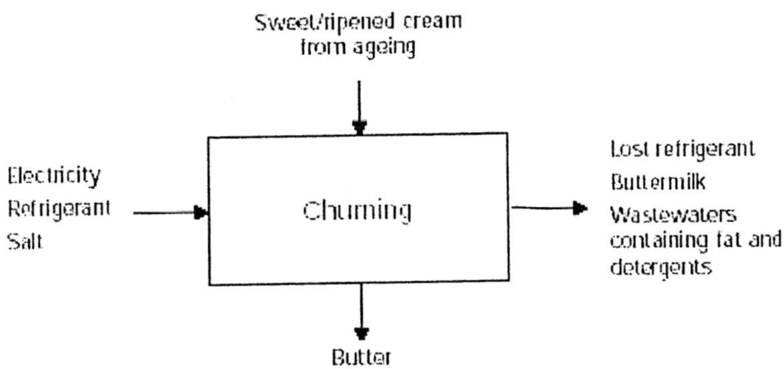

Figure 8. Inputs and outputs of the churning process

Butter may be discharged from the butter maker directly into the feed hopper of a bulk butter packer. However, it is more commonly discharged to a butter silo fitted with a pump, thus avoiding any discontinuities in production. From the silo, butter is pumped to the packing machines through pressure compensators, which control the shear forces.

Butter can be packed initially into 25 kg cases, and subsequently repacked into consumer portions. Alternatively, consumer portions can be packed directly from the continuous butter maker. Most consumer portions are packed in a film wrap (either vegetable parchment or a parchment-lined aluminium foil) or in plastic tubs, which are becoming increasingly popular. Repackaging of bulk butter into consumer portions requires that the frozen butter first be allowed to reach an optimum temperature of 6-8°C, under controlled humidity conditions to avoid excessive condensation. Heat for this process can be provided by carrying out the first stage of thawing in a packed butter store, or low-grade heat from recovery processes.

After temperature adjustment and before repackaging, the butter is reblended to break down the matrix of fat crystals and to re-introduce plasticity. At this stage, there is the opportunity to adjust salt and moisture content to the maximum permitted by local regulation. For repackaging in large quantities, continuous butter blenders are available which incorporate all functions of chopping and blending and prod for inline addition of salt, water and culture. Their construction is similar to that of a continuous buttermaker.

The main environmental issue associated with this process are the high organic loads in the wastewaters generated from rinsing and cleaning the equipment. The greatest potential for environmental loading occurs when machines, such as a continuous butter blender or packing machines, are closed down, because of the residual fat they contain. In addition, product loss may occur when packaged products containing product residues are discarded.

Cleaner Production opportunities in this area focus on reducing water use and loss of product. Ways of achieving this include:

— collecting first rinses while still warm and separating the milk fat residues for use in other processes;

— reducing the disposal of packaging material by having personnel constantly optimising operation of the packaging machines.

Bulk-packed butter is a relatively stable commodity at low temperatures. Commercial freezing stores operate at temperatures down to -30°C, at which temperature the butter should remain in satisfactory condition for more than one year. If storage periods of more than one year are necessary, or if low-temperature refrigerated storage cannot be guaranteed throughout the entire storage life of the butter, the butter can dehydrate below the optimum moisture content. Factors affecting the ability of butter to withstand long-term storage include:

— the cleanliness and hygiene of butter-making operations at all stages;

— the prevention of post-pasteurisation contamination during the addition of salt, moisture etc., and in particular the absence of micro-organisms that grow at low temperatures in the water used for these purposes;

— the degree to which salt (if added) is dispersed to the aqueous phase;

— the overall quality of the butter in terms of its homogeneity, texture and moisture distribution;

— the type of butter.

In order to standardise its consistency and appearance, butter for immediate consumption is placed in cold storage at 5°C for 24-48 hours. This ensures that fat crystallisation is complete and individual packs are firm enough to withstand subsequent transportation to the market. For long-term storage, butter freezing facilities must operate at below -15°C, and temperatures down to -30°C are not uncommon. Sufficient space should be allowed between cases and pallets to allow air circulation, which encourages even chilling.

Refrigeration is usually, but not always, provided by direct expansion ammonia evaporators. Chambers are normally equipped with fork-lift truck access doors protected with automatic door or curtain openers. The main environmental issue associated with the storage of butter is the energy consumed for refrigeration and the potential loss of refrigerant to the atmosphere. Cleaner Production opportunities in this area focus on improving the energy efficiency of refrigerated storage. Ways of achieving this include:

— installing insulation;
— keeping doors closed in cold areas;
— undertaking regular defrosting of cold rooms and regular maintenance of refrigeration systems;
— avoiding refrigerants that contain CFCs. Refrigeration systems based on ammonia cooling are preferred.

Cheese making is an art that is more than 5000 years old. The predominantly rural character of everyday life in the past contributed to the evolution of thousands of different types of cheese and each village, or even family, may have had its own variety, some soft and short lived, some harder and more durable. Modern cheese technology was founded in the nineteenth century when Joseph Harding perceived a need to adopt strict hygiene and control over methods of making cheddar cheese. This represented a step forward in the scientific approach to cheese making.

Cheeses can be categorised according to the following attributes:

— fat content (high-fat, semi-fat and low-fat cheeses);
— consistency (soft cheeses have a moisture content of 45-50% and semi-hard cheese below 40%);
— method of preparation and production.

Some other types of cheese include:

— fresh cheese that can be consumed just after manufacturing and salting (e.g. quark);
— acid-curd cheeses that are coagulated at a higher temperature (e.g. ricotta);
— lactic-curd cheeses which are kneaded or spun (e.g. mozzarella);
— soft cheeses that ripen for only a short time;
— cheeses that develop different tastes due to enzyme action of surface bacteria;
— blue cheeses of many flavours and types;
— semi-hard, mild-tasting, pressed cheeses with holes (e.g. gouda, havarti and tilsit);

— very hard, dry cheeses which are used for grating (e.g. parmesan).

The process description that follows is for the production of cheddar cheese. Cheddar cheese has been used as an example because it is the most widely manufactured and consumed cheese in the world and its industrial manufacturing has been largely automated.

Whole or standardised milk is usually pasteurised at 70°C for 15 seconds and then cooled to the inoculation temperature of 30°C before being poured into a cheese vat fitted with internal agitators. If milk is received on one day and held overnight before being used for cheese production, it will be cooled to 4°C after pasteurisation, and warmed up to inoculation temperature for cheese making.

The starter culture is prepared the day before by the laboratory and may be a single-strain or mixed-strain culture, depending on the flavour required and on the cheese makers' experience. It is important that the mother cultures from which thc daily starter is produced be kept under extremely hygienic conditions in order to avoid contamination-especially from bacteriophages. These are viruses that kill bacteria and can stop cheese-making operations without warning. Each is specific to a bacterial strain, and for this reason the type of starter used is 'rotated' frequently.

Generally, starter is added at the rate of 1-1.5% of the volume of cheese milk. The quantity, however, is determined on a case-by-case basis, depending on starter activity and the subsequent rate of acid development in the cheese milk. When the acidity has reached the required level, usually after 45-60 minutes, rennet is added and dispersed evenly throughout the milk, after which curd formation begins. Rennet acts to coagulate the milk solids into curd. When the curd is firm enough it is carefully cut into cubes the size of large peas. Cutting is done using multiple knives mounted on a frame, which is driven through the curd in two planes.

The mixture of curd pieces and whey is then gradually heated through the walls of the cheese vat to a temperature of 39°C, with slow and careful agitation. Heating assists the process of syneresis, whereby the protein structure shrinks slightly due to the action of the heat, thus expressing whey and creating a firmer curd. During the process of syneresis it is important that the curd pieces not be damaged by the agitators; this could result in a cloudy whey and high losses of fat. When cooking is complete (determined by acidity development and curd structure) the curd pieces are allowed to settle and the whey is drained off. The curd is now one cohesive mass.

Although the process described applies to cheddar cheese, it is similar to that used for other pressed cheese processes. The primary objective is to force whey out of the curd through the action of acidity development, heat and pressure. The curd mass is divided with a knife into blocks. These blocks are turned over and rotated regularly and stacked two or three high. They become thinner as a result of the pressing action. The blocks are kept together as much as possible to maintain warmth. This process continues until the curd texture and the acidity of the whey draining out are at optimum levels. The curd blocks are then milled into pieces about the size of large potato chips.

Dry salt is added to the milled pieces and thoroughly mixed, after which the curd pieces are filled into moulds and pressed overnight. The whey that is expelled from the press station is salt whey and is often white. The moulded blocks of cheese are removed from the moulds and allowed to dry. They are then wrapped in an impervious material-usually a plastic shrink-wrap-and transferred to a ripening room where they remain for about two months, under controlled temperature and humidity, before sale.

Due to the airtight wrapping, maturation during storage is minimal. As a result the majority of cheddar cheeses are mild and bland in flavour and since no rind is formed, all of the cheese can be consumed. So-called 'farmhouse' cheddar cheese is formed and wrapped in a cheesecloth gauze instead of plastic film and is matured for up to six months. This allows the cheese to mature properly and gas to escape slowly, resulting in a product that has fuller flavour, buttery texture and thin rind.

The major environmental issue associated with the cheese-making process is the disposal of whey. There are generally three types of whey:

— sweet whey, which is generated when enzymes, principally rennet, are used to coagulate the milk. Sweet whey typically contains 0.6-0.9% soluble protein, up to 0.3% fat and large quantities of lactose (up to 5%). The pH value of sweet whey from cheddar cheese manufacturing is generally 5.1-5.3;

— acid whey, which is generated when acid is used to coagulate the milk, for example in the production of cottage cheese. Acid whey typically contains the same proportion of soluble proteins as sweet whey, but less fat and somewhat less lactose (4.5%), since some of the lactose is converted to lactic acid. It has a low pH value, between 4.5 and 4.7;

— salt whey, which is the product expressed during the pressing of salted cheese curd, such as in the manufacturing of cheddar cheese. This whey should be collected separately from other types of whey.

Whey produced from natural cheese-making operations contains approximately 6% solids. In the past, whey was perceived merely as an insurmountable problem for the dairy industry because of the high costs of disposal using traditional effluent treatment processes. All too often dairies have taken the easy way out by simply dumping it on land, into rivers or down boreholes. Because of its lactose and protein content, untreated whey has a very high concentration of organic matter which can lead to pollution of rivers and streams and can create bad odours.

A number of opportunities exist for the recovery of the valuable highgrade protein from sweet whey. However it has only been in recent years that they have become technically and economically viable. The method used is ultrafiltration (UF), followed by spray drying of the protein. This process is costly, so is only worthwhile when large quantities of fresh whey are available. Spray-dried whey powder contains between 25% and 80% protein and is used in food products, where it performs a similar function to egg proteins. Whey powder is highly soluble, even at high acidity, and is capable of forming stable foams and gels when heated. Whey protein powder is therefore used in the manufacturing of bakery and meat products, where its gelatinous properties are particularly useful.

Cheese Packaging

After maturation cheeses are packed, either as entire cheeses (in the case of small varieties), or in consumer portions for larger cheeses such as cheddar. Packaging usually involves a combination of manual and automated processes. Packaging materials include natural wax, laminated paper/foil, shrink-wrap plastic, cartons and pre-formed plastic boxes.

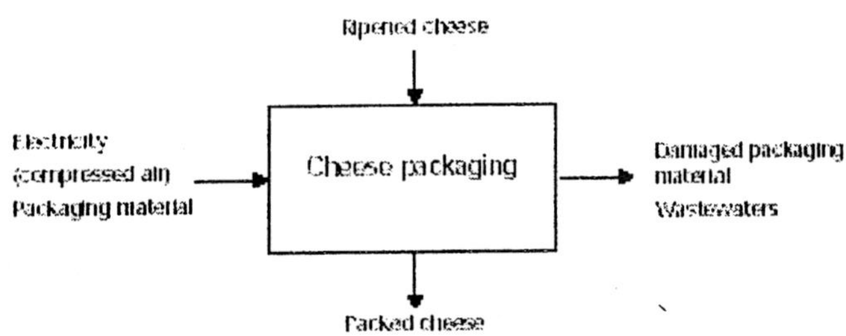

Figure 9. Inputs and outputs for cheese packaging

In some cases it is necessary to clean the surface of the cheese and dry it before packaging. This is most common with cheeses that require a longer maturing time, during which there may be considerable mould growth on the surface. These growths are harmless but not aesthetically pleasing to the consumer.

Cleaning, together with stripping of the cheesecloth bandage, is often a manual process. The process of dividing larger cheeses into smaller portions and then shrink-wrapping them is often a semi-manual operation. Wrapping and boxing of small varieties is normally fully automated. Farmhouse cheddar cheeses, edam, gouda and a few other varieties are often dipped in wax to protect and seal the natural rind. Figure 9 is a flow diagram showing the inputs and outputs for this process.

The major wastes from the cheese packing area are solid wastes, including discarded cuts and small pieces of cheese and damaged packaging material. In addition there are liquid discharges from the cleaning of packaging machines, work surfaces and conveyors. All cheese scraps should be collected separately from other waste and either used as raw material for processed cheese manufacturing (where possible) or sold as animal feed. Liquid wastes should be treated, together with other effluent streams.

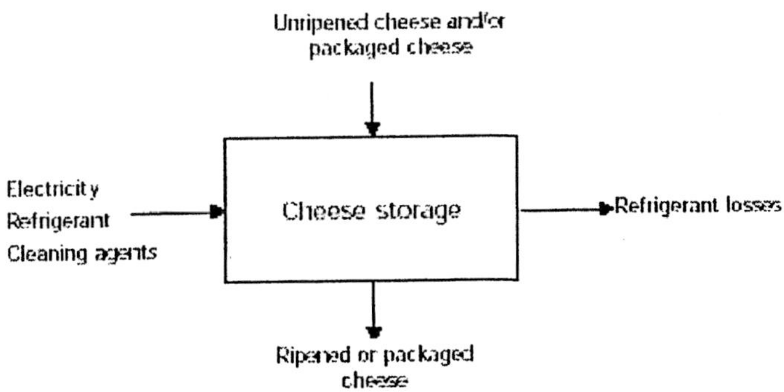

Figure 10. Inputs and outputs for cheese storage

Cheese storage at the processing plant is limited mainly to the ripening period, as cheeses are normally dispatched for sale immediately after final preparation and packing. The temperature of storage varies for different types of cheese. Quickripening soft cheeses require a low temperature of 4.5°C whereas the harder cheeses, requiring longer ripening periods, are normally stored at up to 18°C.

The most important aspect of cheese storage during the ripening stage is humidity control. Humidity may vary from

75% to 85% for hard, dryrind cheeses (such as farmhouse cheddar) to over 90% for soft, rindless cheese or surface-ripened soft cheeses. Figure 10 is a flow diagram showing the inputs and outputs from this process.

The main environmental issues associated with cheese storage are the energy and refrigerants consumed in refrigerated cold stores. Methods for reducing energy consumption and minimising the impacts of refrigerant use are:

— installing good insulation;

— keeping doors to cold rooms closed;

— undertaking regular defrosting and maintenance of refrigeration systems;

— avoiding refrigerants that contain CFCs. Refrigeration systems based on ammonia cooling are preferred.

References

EPRI, *Dairy production in Selected Countries*, Internet document, Palo Alto, CA, USA, 1996.

Farrall, A. W. *Engineering for Dairy and Food Products*, rev. ed., 1980 - P. F. Fellows, Food Processing Technology, 1991.

Harper, W. J., and C. W. Hall, *Dairy Technology and Engineering*, 1981

Hsu, D. A., *Ultra-high Temperature Processing and Aseptic Packaging of Dairy Products*, 1970.

Hui (ed.), Y. H. *Dairy Science and Technology Handbook*, 3 vols., 1994 nd Technology Handbook, 3 vols., 1994.

9

Engineering Properties of Foods

In today's technological world, the meaning of the term has expanded to include not only such disciplines or activities as chemical, medical, polymer, or food engineering, but also genetic engineering and social engineering. Although these disciplines have little to do with engines, they heavily rely on the ingenuity from which the term was originally conceived. It is difficult to define what exactly constitutes an engineering property of a certain food. In general, however, any attribute affecting the processing or handling of a food can be defined as an engineering property. Since many properties are related, there is usually an arbitrary element in their classification. Traditionally, they are divided into the following categories:

- Thermal properties such as specific heat, conductivity, diffusivity, and boiling point rise, freezing point depression.
- Optical properties, primarily colour, but also gloss and translucency.
- Electrical properties, primarily conductivity and permittivity.
- Structural and geometrical properties such as density, particle size, shape, porosity, surface roughness, and cellularity.
- Mechanical properties such as textural and rheological properties (such as viscosity).
- Others, including mass transfer related properties

(diffusivity, permeability), surface tension, cloud stability, gelling ability, and radiation absorbance.

Nearly all of the above properties are manifestations of a food's chemical composition and structural organisation over several orders of length scales-from the molecular to the macroscopic. A change in either composition or structure usually results in a simultaneous change in several properties. Hence it is difficult, if not impossible, to control a single property in isolation.

Food materials or biological materials in general can display large compositional variations, inhomogeneities, and anisotropic structures. Composition can change due to seasonal variations and/or environmental conditions, or in the case of processed foods, properties can be affected by process conditions and material history. For example, North Atlantic fish show dramatic compositional changes in their protein and moisture contents throughout the seasons. Cereals that are puffed up under different moisture and temperature conditions can vary widely in density and .cell-size distribution, and exposure of such products to moist atmospheres, sometimes for short periods only, can have dramatic effects on their crispness.

Therefore, in many cases the data found in published lists for engineering properties of foods can only be considered as approximate values. Nevertheless, these tabular values are still very useful since a safety factor is added to almost all calculations or designs of food processes and/or operations. An understanding of what affects the engineering properties of foods is essential for their proper interpretation and successful utilisation. Therefore, one should always pay attention to the conditions under which the reported properties were determined, especially when response properties are involved. Early physical property analyses of food products required constant uniform values and were often oversimplified and inaccurate. Nowadays, computational engineering techniques, such as the finite element method, are much more sophisticated and can be used to evaluate non-uniform properties that change with time, temperature, and location in food products that are heated or cooled.

Improvements measuring the compositions of foods are now allowing predictions of engineering properties that are more accurate than previously, since they can be predicted from existing numerical and empirical models of the food's composition, temperature, and porosity. There has always been a tendency to make general correlations in predicting properties of food materials for use in process design equations. A myriad of mathematical functions have already been fitted to experimental data, and models are bringing order to experience with the goal of clarifying which components or interactions are important in a food system.

The Engineering Properties of Foods topic covers different sets of engineering properties that are described in greater detail in specific articles, each with wide applications to food engineering and useful for product characterisation and equipment design in food manufacture. Basic definitions, common methods, parameter dependence, modelling, and food engineering applications will dictate the basic pattern followed within most sections.

Thermophysical Properties

Most processed and fresh foods receive some type of heating or cooling during handling or manufacturing. Design and operation of processes that involve heat transfer require special attention due to the heat-sensitivity of foods. Thermal properties of foods are related to heat transfer control in specified foods and can be classified as thermodynamic properties and heat transport properties. Thermophysical properties not only include thermodynamic and heat transport properties, but also other physical properties involved in the transfer of heat, such as freeze and boiling point, mass, density, porosity, and viscosity. These properties play an important role in the design and prediction of heat transfer operations during the handling, processing, canning, storing, and distribution of foods.

Heat can be transferred three different ways: by radiation, conduction, or convection.

- Radiation is the transfer of heat by electromagnetic waves (as in a microwave oven).
- Conduction is the transfer of thermal energy due to molecular oscillations.
- Convection is the transfer of heat by bulk movement of molecules in heated fluids such as liquids or gases.

Although all three types of heat transfer can take place simultaneously, generally only one is predominant, depending on the state of the food and the heating system. In many heat transfer processes associated with storage and processing, heat is conducted through the product; heat is transferred by forced convection between the product and a moving fluid, which surrounds or comes in contact with the product. The thermal properties of foods can characterise heat transfer mechanisms in different unit operations involving heating or cooling. Specific heat, thermal conductivity, thermal diffusivity, boiling point rise, and freezing point elevation are defined as follows:

- Specific heat, C_p, is the amount of heat needed to raise the temperature of unit mass by unit degree at a given temperature. The SI units for C_p are therefore (kJ kg^{-1} K^{-1}). Specific heat of solids and liquids depends upon temperature but is generally not sensitive to pressure. It is common to use the constant pressure specific heat, C_p, which thermodynamically represents the change in enthalpy H (kJ Kg^{-1}) for a given change in temperature *T* when it occurs at constant pressure

$$C_p = (\delta H / \delta T)_p \qquad (1)$$

Only with gasses is it necessary to distinguish between C_p and C_v, the specific heat at a constant volume. Assuming there is no phase change, the amount of heat *Q* that must be added to a unit mass M (kg of mass or specific weight kg/m^3) to raise the temperature from T_2 to T_1 can be calculated using the following equation:

$$Q = MC_p(T_2 - T_1) \quad (2)$$

Thermal conductivity, k, represents the quantity of heat Q that flows per unit time through a food of unit thickness and unit area having unit temperature difference between faces; SI units for are [W m^{-1} K^{-1}]. The rate of heat flow through a material by conduction can be predicted by Fourier's law of heat conduction. A simplified approximation follows:

$$Q = kA(T_1 - T_2)/x \quad (3)$$

where A is the surface area of the food, x is its thickness, T_1 is the temperature at the outer surface where heat is absorbed, and T_2 is the temperature at the inner surface. In other words, κ represents the ability of the food to transmit heat. Unlike specific heat, κ depends on mass density. Thermal diffusivity, α, SI units [m^2/s], defines the rate at which heat diffuses by conduction through a food composite, and is related to κ and C_p through density ρ [kg/ m^3] as follows:

$$a = \kappa / pC_p \quad (4)$$

Thermal diffusivity determines the speed of heat of three-dimensional propagation or diffusion through the material. It is represented by the rate at which temperature changes in a certain volume of food material, while transient heat is conducted through it in a certain direction in or out of the material. Eq. (4) shows that α is directly proportional to the thermal conductivity at a given density and specific heat. Physically, it relates the ability of the material to conduct heat to 'its ability to store heat. In liquid foods, boiling refers to water evaporation, in which water changes from the liquid phase to steam or vapour phase, and water vapour pressure equals the external pressure. Liquid foods contain high molecular weight solids that cause the boiling point to be elevated above that of pure water. The boiling point rise, ΔTr, is known as the increase in boiling point over that of water in a given liquid food. As the vapour pressure of most aqueous solutions is lower than that of water at the same temperature, the boiling temperature (boiling point) of the solution is higher than that of pure water.

During freezing, water in the food changes to ice while heat is removed by a refrigeration system. During heat removal, the unfrozen water will still contain dissolved food solids. The presence of dissolved solids will depress the initial freezing point a certain amount ΔTf below the expected solidification temperature for pure water. Freezing point depression is defined as the temperature reduction ΔTf. Both the boiling point rise and the freezing point depression of a food are related to its solutes concentration.

Determining Thermal Properties Variations

Precision and accuracy of measurement are important factors in determining thermal properties variations. In commercial heating or cooling applications, computer techniques nowadays provide accuracies of 2-5 percent for most heat-transfer calculations, which provide much lower relative errors than practical boundary condition determinations. Several methods are known for measuring specific heat and C_p and thermal conductivity experimentally. C_p measurement of foods can be determined by methods of mixtures and differential scanning calorimetry (DSC). For methods of mixtures, a calorimeter of known specific heat is used and C_p is determined from a heat exchange balance.

In the DSC method, the sample is put in a special cell where the temperature is increased at a constant heating rate. The specific heat of the food is obtained from a single heat thermogram, which relates heat flow as a function of time or temperature. Two experimental methods to determine κ are the Fitch method and the line source method. In the Fitch method, a solid slab of a certain food receives heat from one layer and conducts it to a copper plug. Conductivity κ is obtained from the food's temperature as a function of heat conduction time. The line source method is based on the use of a thermal conductivity probe to measure a temperature-time relation on a thin cylindrical food piece to which constant heat is applied.

Thermal diffusivity α is usually either determined by direct experimental methods or estimated through Eq. (4). Several direct methods for a determination can be based on a one-dimensional heat conduction equation where geometrical boundary conditions are defined. For instance, an apparatus can be used where the sample is located in a special cylinder and immersed in a water bath at constant temperature. Thermocouples located at the center of the sample (axis) and surface of cylinder measure temperature at different heating times. Transient temperature variations are used for the analytical solution. Indirect methods, although they might yield more accurate diffusivity values, require more time and instrumentation for the three-parameter determination (ρ, κ and C_p).

Boiling point elevation ΔTr at a certain external pressure can be determined from a thermodynamic equation using the latent heat of vaporisation and molar fraction of the food. However, the use of these equations requires knowledge of the proportions of specific components of the foods that cause changes in the boiling points. In many cases, estimates for specific components present in higher concentrations can be used. Sometimes reference liquids under the same vapour pressure conditions can be compared with the food, and charts can be used to determining boiling points at different saturation concentrations.

Foods show extended variability in composition and structure, and can be turned into even more complex composite materials when heated together, as in the case of many canned and packed foods, pastry, confectionery, and a wide variety of prepared foods. Thermophysical properties depend on the chemical composition of the structure, determined by the physical arrangement and phase distribution of a system. Thus, heat transfer by conduction may take place in several forms depending on the tortuosity of the material, which may vary at different locations. As porous materials contain a gaseous phase, the value of the thermal conductivity κ, specific heat C_p, and thermal diffusivity α will depend on the internal and external pore space represented by its porosity. Thermophysical properties are significantly influenced by changes in water content and temperature.

During drying, the transfer of heat into food products is accompanied by simultaneous diffusion of water through the product to the surrounding air, provoking differences in thermophysical properties at different regions of the food. Pore size and distribution not only affect heat transfer because of air retention, but also because of the affinity pores have to retain water. The smaller the pore diameter, the greater the surface tension forces, and the more affinity it has for water. Specific heat C_p of foods is drastically influenced by water content.

As a result of solute water interactions, the C_p of each individual component in a food differs from the C_p of a pure component, and usually changes with the concentration of soluble solids. The same occurs with thermal conductivity κ, where water shows greater relative magnitudes in comparison to other food constituents. Thus, both κ and C_p increase with increased moisture content. It is common to find a linear relation between thermal conductivity and moisture content at ambient conditions. The effect of temperature on thermophysical properties is not easy to establish because solids (or semisolids), liquid foods, and food emulsions undergo structural changes. Thermophysical properties of foods change dramatically during the freezing process.

Specific heat changes are difficult to predict when free water becomes solid. Bound water or unfrozen water has a different C_p than bulk-frozen water, and ice has a C_p of about one-half that of liquid water. Thus, C_p below freezing is approximately half that of C_p above freezing. Continuous changes in the fraction of frozen water as temperature varies below the freezing point explain this similarity. In fact, specific heat can be utilized to predict the state of water in frozen foods. Thermal conductivity, however, has been found to be high when temperatures allow water to be in liquid or solid state at very low or high temperatures. Yet when temperatures are within the range of -10 ° to 0 °C, κ shows its lowest values.

Freezing point depression has been modelled with the initial freezing point as a function of water content using linear and quadratic equations. Some thermophysical property models for food systems have been developed as a function of water content or temperature.

General correlations also predict thermal conductivity κ, of food materials for use in process design equations. Linear, quadratic, and multiple correlations of moisture, temperature, and composition can be found for κ in food materials. Some models consider that different components of foods are arranged in layers either parallel or perpendicular to the heat flow. In products such as meats, heat is usually transferred parallel to fibers and κ is dependent on the direction of the heat flow. More general in nature are the randomly distributed models, which consider that the food is composed of a continuous phase with a discontinuous phase dispersed within.

In porous materials, porosity must be included in the model because air has a κ much lower than that of other food components. Models including density or porosity, and pressure, have been developed in fruits and vegetables, meat and meat products, dairy products, cereals, and starch. Several models for predicting a in foods have also appeared in literature; however, most are product specific and a function of water content or temperature. Although the influence of carbohydrates, proteins, fat, and ash on thermal diffusivity has been also investigated, it was found that temperature and water content are the major factors affecting α. Above freezing temperatures, diffusivity varies linearly with temperature or water composition in some foods, while this is not valid at below-freezing temperatures.

Food Processing Applications

Food thermal properties play an important role in the quantitative analysis of food processing operations. Numerous food processing unit operations are heat or energy sensitive. Heat exchanged and resulting temperature-pressure relations must consider the minimisation of reactions such as browning, vitamin loss, and oxidation reduction in order to preserve the acceptability and nutritional value of the food. Thermal properties are useful when evaluating capacities of drying systems, or studying the effect of product shrinkage or internal cracking with the aid of mathematical and numerical drying models. Enthalpy and specific heat are required to calculate the heat load in food processing operations. Specific heat measurement allows evaluation of the structure of foods.

During processing involving heat, temperature within a food changes continuously, varying not only the food C_p but also the κ. When conduction of heat is involved, thermal conductivity is important to predict or control the heat flux and processing times. In a processing system, it is necessary to predict the time end-point of processing to ensure the efficiency of the equipment. It is also desirable to heat and cool foods as rapidly as possible to improve the economics of the process by increasing the capacity and delivering a better quality product. All processing-time prediction models need the thermal conductivity data of food where energy transfer is involved. The speed of heat propagation or diffusion through the material is also related to processing times. Therefore, thermal diffusivity can also participate in processing-time estimation of processes like canning, cooling, freezing, and frying.

The equilibrium freezing point can be used for the prediction of thermophysical properties because of the discontinuity exhibited at that point. Accurate freezing point data can also be used to calculate other colligative properties such as effective molecular weight, water activity, bound, free, and frozen water, and enthalpy below freezing point. Knowledge of freezing point is important for analysing freezing and thawing times of frozen foods. Freezing point data can be used to ascertain chemical purity with regard to whether a sample differs from a natural or desired condition.

The increase in boiling point or boiling point rise (ΔT_r) of liquid foods is a property of interest in evaporators or other types of heat exchanger equipment design and operation. It is worth mentioning the role of the surface heat transfer coefficient, as it is one of the important parameters necessary to design and control food processing equipment where fluids participate. Although it is not a property of a food, it is used to quantify the transfer rate of heat by convection from a liquid or a gas to the surface of the foods. It plays an important role when evaluating the effectiveness of heat transfer in processes where hot water or steam is applied through the evaluation of the overall resistances during heat transfer.

Optical Properties

Optical properties are those material properties resulting from physical phenomena occurring when any form of light interacts with the material under consideration. In the case of foods, the main optical property considered by consumers in evaluating quality is colour, followed by gloss and translucency or turbidity among other properties. "Colour" is the general name applied to all sensations arising from the activity of the retina, and is related to visual appearance of food (shape, size, surface and flesh structure, and defects).

Optical properties are related to consumer judgment on food appearance and produce some kind of visual effect. Among these, colour, gloss and translucency can be defined as follows.

- Gloss is the name given to light specularly reflected from a plain smooth surface. It can be defined by a goniophotometric curve, which represents the intensity of light reflected at the surface at different angles of incidence and viewing.
- Colour is essentially a beam of light composed of irregularly distributed energy emitted at different wavelengths. Depending on the type of illumination, the same material can show different light qualities and produce different sensations. Foods, along with other materials, have colour properties, which depend exclusively on their composition and structure.
- Translucency of foods is defined using an opaque-to-transparent scale. In liquid foods, light passing through

changes its path randomly when interacting with suspended particles. Although light can be transmitted or reflected, the human eye only experiences translucency as a sensory attribute distinct from colour. Many food products are neither fully opaque nor fully transparent, but are translucent.

The colour perceived when the eye views a food is related to the following three factors: the spectral composition of the light source, the chemical and physical characteristics of the food, and the spectral sensitivity properties of the eye. To evaluate the colorimetric properties of a food, two of these factors must be standardised. Although the human eye can give fairly uniform results, it can be replaced by some instrumental sensor or photocell to provide even more consistent determinations. Visual colorimeters facilitate visual comparisons and eliminate differences in interpretation between operators.

In practice, visual measurement of colour entails comparing the colour problem with reference colors available in printed charts under well-defined and favourably conditions for good, reproducible comparisons. Light source, geometry of viewing, and colour of background are the most important factors to control. Description of colour for purchase specifications of food commodities or packaging materials involves colour tolerances, which are defined in one, two, or three dimensions in colour space to avoid variability of the human eye. Several systems of colour analysis have been created. The most used are the CIE, Munsell, Hunter, and Lovibond systems.

— In the CIE system, spectral curves indicate how the eyes of normal observers respond to various spectral light types in the visible portion of the spectrum. The system is based on the fact that any colour can be matched as a suitable mixture of red, green, and blue. These primary combinations are called tristimulus values of colour. A certain colour can be defined by chromacity coordinates x and y, and by the luminous transmittance or lightness. A chromacity diagram defines different colour points that define the standard colour of a food. The US Department of Agriculture uses chromacity coordinates to define specifications of colour standards for a variety of products.

— In the Munsell system, all colors are described by three attributes: hue (or type of colour), lightness, and saturation or purity (associated with clear to dark perception). The hue scale is based on ten hues distributed on a circumference (scaled 1 to 10); the lightness ranges from black to white (0 to 10) and is distributed on a perpendicular line; the purity is of irregular length beginning with 0 for the central gray to the limit of purity obtainable by available pigments in the Munsell book of colour. The Hunter system is also a three-dimensional system using parameters L*, a*, and b* in each dimension: L* is the lightness (nonlinear), a* is redness or greenness, and b* is yellowness or blueness. Combination of L*, a*, and b* can be converted to a single colour.

— The Lovibond system is a standard method generally used to determine the colour of vegetable oils. It involves visual comparisons of light transmitted through a glass couvette using colour filters. Vegetable oils are usually expressed in terms of red to yellow. The Lovibond index can also be used to measure colour in wines and juices. Computer software packages have been developed that easily convert light transition spectra into CIE, Munsell, Hunter, and Lovibond colour indices.

Colour can be measured instrumentally with colorimeters, which may be broadly classified as tristimulus colorimeters and spectrophotometers. The difference between spectrophotometers and colorimeters is that the former measures intensity of light through the completely visible spectrum, and colorimeters are designed to measure only some parameters related to sensory colors. Colorimeters are very useful in the quality control of foods, and give results normally correlated with visual measurements.

A Munsell colorimeter consists of a circular rotating platform where several coloured disks are mixed in different proportions to provide a range of shades to match the colour of a certain food product. It is widely used in the food industry for quality control of a number of solid products like tomatoes, fruits, and peanut butter. Tristimulus colorimeters measure both related scales of Munster, Hunter, or CIE systems, which are numerically related. The quality of output for this type of instrument mainly depends on the correct combination of light source, filter, and photocell to obtain a good reproduction of visual response.

Glossmeters measure intensity of light reflected at three angles of incidence and reflection, and normally give results in the form of indices, obtained by comparing the sample reflectance to that of a highly reflective flat glass, used as a calibration standard. These indices are easy to interpret, in contrast to more difficult goniophotometric curves used in the past for classification. Translucency can be the measurement of the reflection of a thin sample against both a white and black background. From these measurements, the value of reflection from an opaque layer is calculated as a ratio between absorption and scattering to measure scattered light.

Glossiness of a product is a property of the smoothness of its surface. When this characteristic is desired, manufacturers try to improve it, as in the case of fruits covered with wax to make them more visually appealing. Translucency is also worth consideration in some liquid foods, such as fruit juices. Its measurement can be determined by considering the contributions of both absorbed and scattered light when traversing these products. For a few clear liquid foods, such as oils and beverages, colour is mainly a matter of transmission of light. Other foods are opaque and derive their colour mostly from reflection.

Optical properties can be used to perform quality control and continuous inspection during processing operations. Major requirements for a quality control system are ease of calibration and use, stability, precision, speed, cheapness, and industry-wide applicability. A complete colour description requires the use of three dimensions, and a control automatic system may be based on this complete specification. Specifications may be set to provide an idea of fruit ripeness, milk or cream discoloration during sterilisation, degree of roasting of coffee grains, or browning of apples slices during storage.

Continuous colour measurements are used in tasks involving colour sorting (or "electronic sorting") by using in-line systems. Colour sorting is used for a very wide range of food materials in screening defects. Visible, infrared, and ultraviolet laser beams can provide continuous inspection through scanning of product size, symmetry, damage, irregular shape, fill level, and label placement by adding automatic software in connection with mechanical devices.

Electrical Properties in Food Engineering

There are two main electrical properties in food engineering: electrical conductivity and electrical permittivity. Electrical properties are important when processing foods involving electric fields, electric current conduction, or heating through electromagnetic waves. These properties are also useful in the detection of processing conditions or the quality of foods. Electrical conductivity is a measure of how well electric current flows through a food of unit cross-sectional area A, unit length L, and resistance R. It is the inverse value of electrical resistivity (measure of resistance to electric flow) and is expressed in SI units S/m in the following relation:

$$\sigma = L/(AR) \quad (5)$$

Electrical permittivity is a dielectric property used to explain interactions of foods with electric fields. It determines the interaction of electromagnetic waves with matter and defines the charge density under an electric field. In solids, liquid, and gases the permittivity depends on two values:

— the dielectric constant ε', related to the capacitance of a substance and its ability to store electrical energy; and

— the dielectric loss factor ε″, related to energy losses when the food is subjected to an alternating electrical field.

The electrical conductivity of foods has been found to increase with temperature (linearly), and with water and ionic content. Mathematical relationships have been developed to predict the electrical conductivity of food materials: for example, for modelling heating rates through electrical conductivity measurements, or for probability distribution of conductivity through liquid-particle mixtures. Electrical conductivity shows different behaviours during ohmic and conventional heating.

At freezing temperatures, electrical conductivity increases with temperature, as ice conducts less well than water. Starch transitions and cell structural changes affect electrical conductivity, and fat content decreases conductivity. As in thermal properties, the porosity of the food plays an important role in the conduction of electrons through the food. In foods, permittivity can be related to chemical composition, physical structure, frequency, and temperature, with moisture content being the dominant factor. Dielectric properties (ε′, ε″) are primarily determined by their chemical composition and, to a much lesser extent, by their physical structure. The influence of water and salt (or ash) content largely depends on the manner in which they are bound or restricted in movement by other food components. Free water and dissociated salts have a high dielectric activity, while bound water-associated salts and colloidal solids have low activity.

Power dissipation is directly related to the dielectric loss factor ε″ and depends on the specific heat of the food, density of the material, and changes in moisture content. Permittivity also depends on the frequency of the applied alternating electric field. Frequency contributes to the polarisation of molecules such as water. In general, dielectric constant increases with temperature, whereas loss factor may either increase or decrease depending on the operating frequency. Both the dielectric constant ε′ and loss factor ε″ decrease significantly as more water freezes. Reasonable comprehensive tabulations of electrical properties data are available for foods in electronic and printed form.

The conductivity of a material is generally measured by passing a known current at constant voltage through a known volume of the material and by determining resistance. The total conductivity is then calculated simply by taking the inverse of the total resistivity. Basic measurements involve bridge networks (such as the Wheatstone bridge circuit) or a galvanometer. There are other devices that measure electrical conductivity of foods under ohmic or conventional heating conditions, using thermocouples and voltage and current transducers to measure voltage across and current through the samples.

Known methods for measuring dielectric properties are the cavity perturbation, openended coaxial probe, and transmission line methods. Since modern microwave network analysers have become available, the methods of obtaining dielectric properties over with frequency ranges have become more efficient. Computer control of impedance analysers and network analysers has facilitated the automatic measurement of dielectric properties over wide frequency ranges, and special calibration methods have also been developed to eliminate errors caused by unknown reflections in the coaxial-line systems. Distribution functions can be used in expressing the temperature dependence of dielectric properties.

Electrical properties are important in processing foods with pulsed electric fields, ohmic heating, induction heating, radio frequency, and microwave heating. Conductivity plays a fundamental role in ohmic heating, in which electricity is transformed to thermal energy when an alternating current (a.c.) flows through food. As it has potential use in fluid pasteurisation, it is important to know the effective conductivity or the overall resistance of liquid-particle mixtures.

The electrical field inside the food is determined by the dielectric properties and the geometry of the load, and by the oven configuration. These properties are also useful in detection processing conditions, or the quality of foods. The major uses for dielectric properties are measuring and heating applications. Permittivity and moisture are closely correlated when the water content is high. Properly designed electrical instruments can be used to determined moisture content or water activity.

Knowledge of dielectric properties in partially frozen material is critical in determining the rates and uniformity of heating in microwave thawing. As the ice in the material melts, absorption of energy increases tremendously. Thus, the portions of material that thaw first absorb significantly more energy and heat at increasing rates, which can lead to localised boiling temperatures while other areas are still frozen. Salt affects the situation through freezing point depression, leaving more water unfrozen at a given temperature.

Mechanical Properties

The mechanical properties mainly result from the structure, physical state, and rheology. They can be subdivided into two groups: structural and geometrical properties, and strength properties. Structural and geometrical properties include mass-volume-arearelated properties, and morphological properties. Strength properties are related to solid and semi-solid stress and deformation, and intervene in food texture and rheological characterisation. These properties are needed for process design, estimating other properties, characterising foods, and quality determination.

Structural and Geometrical Properties

Forms of Density

This is defined as mass per unit volume (the SI unit of density is kg/m^3). Indeed, there are different forms of density such as true, material, particle, apparent, and bulk that can be used, depending on its application in process calculations or product characterisation. The volume measurement method is what determines the difference between them. True and material densities are calculated by excluding volumes occupied by internal and external pores within the food, while particle, apparent, and bulk densities are determined from less accurate measurement methods that include pore volume.

A material's volume can be measured by buoyant force, liquid, gas or solid displacement, or gas adsorption, or by estimating the material's geometric dimensions. The buoyant force method for apparent or particle volume determination utilises sample weight differences in air and water, while the liquid displacement method measures the increase in liquid volume when the material is immersed in a non-wetting fluid such as mercury or toluene. A gas pycnometer is a gas displacement device that uses highpressure air differences in a sample cell connected to a manometer to determine material volume. Apparent or particle density can be determined by coating particles in order to include internal pores in the volume measured. For solid displacement, sand or glass beads can be used instead.

In most engineering designs, solids and liquids are assumed to be incompressible-in other words, density changes moderately with changes in temperature and pressure. In food engineering, the density of solid and liquid foods changes with temperature and pressure and is dependent on temperature and composition. In the case of liquid foods, no generic equations exist to predict the density. In the literature most of the density data is correlated empirically as a function of temperature, water, solids, and fat content. Different types of nonlinear correlation, such as exponential, quadratic, and cubic, are used to relate density and moisture content.

Porosity

Porosity indicates the volume fraction of void space or air space inside a material. Volume determination is relative to the amount of internal (or closed) or external (or open) pores present in the food structure. Therefore, like density, different forms of porosity are also used in food processing studies, namely open pore, closed pore, apparent, bulk, and total porosities. Porosity can be measured by direct and microscopic methods, or can be estimated from density data. Porosity in foods is mainly predicted from empirical correlations, which are valid for individual foods under given processing conditions. Fundamental models exist that are based on the conservation of mass and volume, as well as a number of terms that account for interaction of components and formation or collapse of air or void phase during processing.

Shrinkage

This is the reduction in volume or geometric dimensions during processing. When postprocessing volume is larger than initial volume, it is termed as expansion. Two types of shrinkage-isotropic and anisotropic-are usually observed in the case of food materials. Isotropic shrinkage is described as the uniform shrinkage of the materials under all geometric dimensions, whereas anisotropic (or non-uniform) shrinkage develops in different geometric dimensions. The former is common in fruits and vegetables while the latter is known in animal tissue, such as in fish.

Shrinkage occurs as a result of moisture loss, ice formation (during freezing), and formation of pores . The glass transition theory is one of the concepts proposed to explain the process of shrinkage, collapse, fissuring, and cracking during drying. The methods of freeze-drying and hot-air drying can be compared on the basis of this theory. Pore disruption and structure hardening, as well as moisture transport mechanisms, counterbalancing internal forces, and environmental pressure, are some of the causes for reduction or collapse in a food structure.

Expansion can be caused by gas generation, which is mainly a result of water evaporation and subsequent pore formation within the food structure. More work is needed to develop a fundamental understanding of how pores are formed, or of the collapse mechanism during processing, and their impact on product characterisation. Most of the density, shrinkage, and porosity prediction models for liquid and solid foods are empirical in nature. Recent models have been developed to predict porosity during air-drying based on drying temperature, moisture content, initial porosity, and product type.

Two types of surface areas are used in process calculations and product characterisation-the outer or boundary surface of a particle or object, and the total surface area of a porous object, or the pore surface area. It is very common to estimate the surface area from its geometric dimensions in the case of a Euclidian geometric object. A Euclidian geometry always has characteristic dimensions and assumes surfaces to be smooth, such as in spheric, cubic, and ellipsoidal geometries. Many natural patterns are either irregular or fragmented, to such an extreme degree that Euclidian or classical geometry is no help in describing their form.

Fractal analysis is used instead to characterise and estimate the surface areas of these shapes. Native and physically or chemically transformed food particles can be characterised by fractals to predict the efficiency of the transformation process and food particle properties, such as adsorption capacity, solubility, puffing ability, chemical reactivity, and emulsifying ability to optimise food ingredient selection for product development and process design. Like fractal analysis, neural networks or artificial intelligence may have potential in modelling surface, shape, and other mechanical properties of foods.

Morphological properties such as roundness and sphericity are also used to characterise a food's shape. Roundness is a measure of the sharpness of the corners of a solid. Sphericity indicates how the shape of an object deviates from a sphere. Sphericity is defined from the volume, surface area, or geometric dimensions of an object. Sphericity and shape factors are also needed in heat and mass transfer calculations. Size, shape, sphericity, volume, surface area, density, and porosity are important physical characteristics of many food materials in handling and processing operations. Fruits and vegetables are usually graded according to size, shape, and density.

Impurities in food materials can be separated by density differences between impurities and foods. Values for surface areas of fruits and vegetables are needed in investigations related to respiration rate in heat transfer calculations for heating or cooling. Density and the shape factor of food materials are also necessary for predicting the freezing and thawing rate. Volume change and porosity are important parameters in estimating diffusion coefficients for shrinking systems. Porosity and tortuousity are used to calculate effective diffusivity during the mass transfer process.

Rheology and Texture

Mechanical properties are intertwined with rheology when including strength properties. Mohsenin defines mechanical properties as "those having to do with the behaviour of the material under applied forces." Rheology has been defined as "a science devoted to the study of deformation and flow," or as "the study of those materials that govern the relationship between stress and strain." "Stress" is defined as force components acting on a body per unit cross-sectional area or area of the deformed specimen (SI units in Pa). "Strain" is the change in size or shape of a body in response to the applied force. Rheologically, the behaviour of a material is expressed in terms of stress, strain, and time effects. Therefore, properties that deal with the motion of the material as a result of an applied force can be included as mechanical forces.

Rheological tests express stress-strain relationships and study strain rate dependency. Ideal solids deform in an elastic Hookean manner, while ideal liquids flow in a viscous Newtonian manner; in each case the behaviour is independent of the strain rate. Nonetheless, foods are strain-rate dependent. They usually contain some solid and liquid attributes and, rheologically, are termed viscoelastic bodies. There are three stresses that are commonly applied to characterise foods mechanically: compressive, tensile, and shearing. Shear stress is the most prevalent with fluids or viscous materials. Since strain is the response of the material to stress, compressive shear and tensile strains can be found. When small deformations are exerted under compression, foods can show a straight line in the stress strain plot, and its slope is called the "Young modulus of elasticity."

Rheologically, a material can deform in three ways: elastic, plastic, or viscous; it can be denoted by a spring friction element and a dashpot arranged in series or parallel, respectively, in rheological models. Different mechanical situations define how stress can act on a food: static, dynamic, or impact. Impact during mechanical handling is the most common cause of mechanical damage to foods. Behaviour under static or dynamic stresses governs the extent of potential mechanical injury and can provide valuable information on the design of handling machinery. In cases like these, definitions of creep or stress relaxation play a role.

Solid foods are mechanically characterised by compression tests or impact tests. Universal testing machines give curves of normal force versus deformation, shear forces, creep, and stress relaxation measurements. The most important mechanical-rheological behaviour of fluid or viscous foods is the flow behaviour, which can be basically defined as Newtonian, pseudoplastic, and Bingham, indicating viscosity of the material and its dependence on shear rate. In processing, flow properties can influence pumping requirements, flow of fluid through pipes, or even extrusion properties. Flow properties can be determined using any variety of available rheometers or viscometers.

The mechanical properties of foods intersect not only with rheological behaviour but also with the texture of foods. In fact, mechanical properties form the basis for food's sensory properties related to texture. For instance, during both mastication and industrial size-reduction processes, it is desirable to weaken the structure so that it will properly disintegrate when forces are applied.

In this way, texture can be defined as those physical characteristics arising from the structural elements of the food that are sensed primarily by the feeling of touch, related to deformation, disintegration, and flow of the food under a force, and measured objectively by functions of mass, time, and length. This indicates that texture studies include structure and the manner in which structure reacts to applied forces. It also emphasizes that texture is a multidimensional property comprising a number of sensory characteristics.

For instance, the Texture Profile Analysis widely used in industry defines mechanical parameters such as hardness, fracturability, cohesiveness, springiness, chewiness, gumminess, and resilience. Compression tests evaluate texture by compressing the food in one direction and unrestraining it in the other two dimensions to evaluate hardness or strength in solid foods. The puncture test is the oldest one used for food texture determination, and involves the measurement of force necessary to penetrate the test material with a punch. Sometimes puncture tests imitate the failure involved in mastication and can measure the firmness of a fresh fruit. The introduction of computer readout and analysis of forcetime plots obtained with Universal Testing Machines allow reading of the maximum forces from the force-time graphs, measuring of slopes, and calculation of areas under the curve, among other features. Highly complex analyses of force-time plots have now become routine.

Structure may refer to the often-complex organisation and interactions of food components under the influence of external and internal physical forces. Structure also refers to the size and shape of the components of the food, as well as how they interact to form an organisation. Thus, texture properties can predict deformation mechanisms after stress application, effects of heating in baked products, thawing or freezing mechanisms in meats, or changes in the hardness of fleshy fruit tissues during ripening, through structural microscopy studies. Surface structure microscopy can complement the characterisation of the strength properties through traditional qualitative methods such as scanning electron microscopy or confocal laser scanning, among other microstructural methods.

References

Aguilera J.M and Stanley D.W., *Microstructural Principles of Food Processing and Engineering*, 2nd Edition, Gaithersburg, MD: Aspen Publishing, 1999.

Bibliography

Archer, G.P., Kennedy, C. J. and Wilson, A. J., "Towards predictive microbiology in frozen food systems-a framework for understanding microbial population dynamics in frozen structures and in freeze-thaw cycles", *International Journal of Food Science and Technology*, 30, pp. 711, 1995.

Blanshard, J. M.V. and Franks, F., *Ice crystallization and its control in frozen-food systems*, Food Structure and behavior, Academic Press, London, UK, pp. 51-65, 1987.

Brady, P.L., *Freezing fruit*, UA Cooperative Extension Service, University of Arkansas, USA, 2002.

Brennan, J. G., Butters, J. R., Cowells, N. D. and Lilly, A. E., *Food Engineering Operations*, 3rd Edition, Applied Science, London, 1990.

Cano, P., Marin, M.A. and Fuster, C., "Freezing of banana slices, Influence of maturity level and thermal treatment prior to freezing", *Journal of Food Science*, 55, 4, pp. 1070, 1990.

Charm, S. E., *The Fundamentals of Food Engineering*, 2nd Edition, AVI, Westport, 1971.

Chen, C., "Thermodynamic analysis of freezing and thawing of foods enthalpy and apparent specific heat", *Journal of Food Science*, 50, 1985.

Coulson, J. M. and Richardson, J. F., *Chemical Engineering*, 3rd Edition, Vols. 1, 2, Pergamon, Oxford, 1977, 1978.

Danny D. Reible., *Fundamentals of Environmental Engineering*, CRC Publishers, 1998.

Dixon, J. A., *Technical Support Services for Agro-Industrial Development-Jamaica*, Paper presented at the Regional Consultation on Agro-Industrial Development, September 28-29, Port-of-Spain, Trinidad and Tobago, 1998.

Emken, E. A., "Nutrition and biochemistry of trans and positional fatty acid isomers in hydrogenated oils," *Annual Reviews of Nutrition*, 4: 339-376, 1984.

Fellows, P. and Hampton, *A Small-Scale Food Processing A Guide to Appropriate Equipment*, London: Intermediate Technology Publications, 1992.

Grabe, D. F. and D. Isely., "Seed storage in moisture-resistant packages" *Seed World*, Jan 24, 1969, p. 4.

Harrington, J.F., "Vegetable seed germination temperatures," *Agricultural Extension leaflet*, University of California, Davis, 1954.

Hunsigi, G., *Agricultural Fibres for Paper Pulp*, Outlook on Agriculture 18 (3): 96-103, 1989.

James R. Mihelcic, Martin T. Auer, and others., *Fundamentals of Environmental Engineering*, John Wiley, 1999.

Jones, G., *The Role of Science and Technology in Developing Countries*, Oxford University Press, London, England, 1971.

Loncin, M. and Merson, R. L. *Food Engineering; Principles and Selected Applications*, Academic, New York, 1979.

Navazio, J., "Seed-born diseases" (lecture handout), *Organic Seed Alliance*, Port Townsend, WA, 2003.

Talbot, M. T. and Fletcher, J.H., *Design and Development of a Portable Forced-Air Cooler*, Proceedings of the Florida State Horticultural Society 106:249-255, 1993.

Tchobanoglous, G., Burton, F.L., and Stensel, H.D., *Wastewater Engineering (Treatment Disposal Reuse) / Metcalf & Eddy, Inc.*, 4th Edition, McGraw-Hill Book Company, 2003.

Turner, D.B., *Workbook of atmospheric dispersion estimates: an introduction to dispersion modeling*, 2nd Edition, CRC Press, 1994.

Index

Agro processing technology 3
Agro-processing programmes 29
All India Coordinated Research Projects (AICRP) 4

Barrel-type washers 126
Blanching treatment 118
Bureau of Indian Standards (BIS) 26

Clostridium botulinum 149
Commercial freezing systems 174
Commercial preservation techniques 173
Computational engineering techniques 246
Computer software packages 257

Differential Scanning Calorimetry (DSC) 250

Frozen food industry 172
Frozen vegetable industry 174

Heat resistant enzymes 118
Heat transfer mechanism 190, 191
Heat transfer systems 188
Hydrochlorofluorocarbon (HCFC) 185

Immersion freezing systems 192
Indigenous technology 17
Individually Quick Frozen (IQF) 191
International Institute of Refrigeration (IIR) 179

Liquid Nitrogen (LIN) 185

Mechanical handling systems 10
Mechanical refrigeration system 185
Metal screen trays 70
Microcrystalline Cellulose (MCC) 19
Million Tonnes (Mt) 3
Mixed pickles 146
Moisture Vapour Transmittance Rate (MVTR) 128
Moisture-vapour-proof 195
Multi-plate freezers 177

Natural acidification technology 139
Natural textile fibre 18

Non-acid foods 150

Opaque packaging materials 124

Poor post harvest management 3
Post harvest management. 2
Post Harvest Technology (PHT) 4
Public distribution system 9

Sauerkraut technology 139

Threshing machines 1
Tomato paste pasteurisation 135

Vegetable blanching treatments 121
Village industries commission 18
Volume measurement method 262

Water-soluble nutrients 203
World frozen-food categories 175